THE (NEW) TURING OMNIBUS

THE (NEW) TURING OMNIBUS

66 Excursions in Computer Science

A. K. Dewdney
The University of Western Ontario

A W. H. Freeman / Holt Paperback
Henry Holt and Company   New York

Holt Paperbacks
Henry Holt and Company, LLC
Publishers since 1866
175 Fifth Avenue
New York, New York 10010
www.henryholt.com

Library of Congress Cataloging-in-Publication Data
Dewdney, A. K.
  The (new) Turing omnibus / A. K. Dewdney.
  p.   cm.
  Includes index.
  ISBN-13: 978-0-8050-7166-5
  ISBN-10: 0-8050-7166-0
  1. Electronic data processing. 2. Computers. I. Title.
QA76.D448   1993   93-17330
004—dc20   CIP

Henry Holt books are available for special promotions and
premiums. For details contact: Director, Special Markets.

First Holt Paperbacks Edition 2001

A W. H. Freeman / Holt Paperback

Printed in the United States of America

D 30 29 28 27 26 25 24 23

Dedicated to the memory of

Assna Bernstein
(1902 – 1969)

German lessons,
endless toast and jam,
the human cabbala

CONTENTS

PREFACE

The (New) Turing Omnibus is a vehicle of learning that visits the major land-marks of computer science. The tour stops at monuments of theory, visits major techniques, and travels the avenues of application. This new edition of *The Turing Omnibus* features a new tour, in effect, with new landmarks added to the itinerary. In the Preface to the 1989 edition, I described the collection as "not an exhaustive survey . . . but largely subjective." With six new chapters and many chapters revised, the subjectivity matters less. And the book is that much closer to being exhaustive, at least at its natural level of exposition.

The metaphor of a *touring* omnibus emerged during a sabbatical taken at Oxford in 1980 – 1981. Large, red double-decker omnibuses roared from stop to stop in the urban Oxford landscape and, farther south, in London. Here was a symbol of England, the land of Alan Turing, whose ripe mathematical imagina-tion was fueled by fantasies of mechanized intellect. Turing's greatest single legacy, the Turing machine itself, became the best-known vehicle for the emerging theory of computability. Wonderfully simple, with only two moving parts, it can be explained to the man or woman in the street. Yet it embraces everything that we mean by the word "computable"! Not surprisingly, this book visits a Turing machine, still in working order.

The Turing Omnibus began during that fateful sabbatical at Oxford. My host was Dominic Welsh, a well-known mathematician of Merton College. Interested in all things computational, Welsh would conduct occasional walks along the Cherwell. One day he bemoaned the lack of a good, comprehensive introduction to computing, not a dry text — there are plenty of those — but a semipopular compendium of computer science, the "good bits," as he called them. "The topics," he said, "must have charm or, at least, relevance." I felt in immediate sympathy with the idea. I too had often wished for something similar, albeit more academically directed. Over succeeding months we put down a list of topics on paper. Welsh, *not* being on sabbatical at the time, gave up the project but encouraged me to continue.

Sometime during my childhood I encountered the traditional image of a bird that erodes a mountain by taking a single stone from it every year. In time, a mountain may disappear by this method — and appear elsewhere. That was my feeling about the incremental growth of *The Turing Omnibus* from 1981 to 1988 when I completed the first draft, astonished once more at the power of incremental work. The topics came not only from the original list drawn up with Welsh, but also from badgering colleagues at my home institution, the University of Western Ontario, and other colleagues met at conferences all over North America during the early 1980s. Since the publication of the 1989 edition, I have badgered even more people and, as a result, developed more topics. Two ideas, the Mandelbrot set and computer viruses, originated in my Computer Recreations column in *Scientific American* magazine.

The result, *The (New) Turing Omnibus*, is actually more than a "tour" and comes, by virtue of its new material, closer to being an "omnibus" in the sense of a complete collection. This book is designed to appeal both to the educated layperson and to the student of computer science. But how is that possible? The answer lies in the variety of treatments as well as topics. Some of the topics are inherently easy or I have been lucky enough to stumble upon just the right expository mechanisms. Some of the topics are inherently deep or complicated and there is no way around a certain rigor, including occasional mathematical symbolism.

For students of computer science, the 66 chapters that follow will give a sneak *preview* of the major ideas and techniques they will encounter in their undergraduate careers and some they may only encounter as graduate students. For professors of computer science, my colleagues, the 66 chapters will amount to a sneak *review*. Trying to remember just how the Boyer-Moore string-matching algorithm went? It's right there in Chapter 61, Searching Strings. As for your lectures, if you like to deliver your own material this book may be what you've been looking for. At one end of its spectrum of uses, *The (New) Turing Omni-*

bus may be ideal in bringing students from diverse backgrounds "up to speed." At the other end of the spectrum, you retain creative control but draw a few (or many) of your lectures from this book. Finally, for educated laypersons, the book provides a brief roadmap of computability.

Inside *The (New) Turing Omnibus*

The (New) Turing Omnibus, like its predecessor, is intended primarily for use as a reference work on the theory and practice of computing for both students and professionals. But, as a number of course adoptions have demonstrated, it also works quite well in a variety of lecture settings, including the theory of computation, analysis of algorithms, computer logic and systems, data structures, and even artificial intelligence. The adoptions were possible only because I took pains to introduce topics that, while not earthshaking in themselves, provided a key step in the development of a particular subject. For example, the logic behind memory addressing (Chapter 38, A Computer Memory) shakes nobody's Earth, but it forms an essential part of the development of ideas from simple circuits to the full-blown random access computer. On the other hand, Welsh might well find a certain charm in the use of encoders and decoders to load parallel bits into and out of memory.

In the new edition, I have augmented the logic and systems stream with two new chapters, one on disk operating systems (Chapter 53) and the other on computer viruses (Chapter 60). I have added a chapter on Newton's method (Chapter 21) because no course in numerical methods is complete without this beautiful use of derivatives. I have added a chapter on genetic algorithms (Chapter 16) as yet another approach to artificial intelligence—but without greatly questioning its validity. The jury is still out on a method that (a) claims to solve difficult problems and (b) is suspiciously painless. The Mandelbrot set is such a magnificent object and only computers were able to show us what it looks like. More formally, it augments the applications area by demonstrating the use of computers in mathematical research.

I have deleted one chapter from the 1989 edition, the one on neural nets and logic. I have added a new chapter on how neural nets learn (Chapter 36, Learning Networks). I was tempted to supplement this by another chapter on how neural nets *don't* learn but decided to keep my own finger out of the pie, to be a compiler rather than a whole operating system. Besides, a chapter retained from the 1989 edition (Chapter 27, Perceptrons) carries an ample warning to those who expect too much from a given computational strategy. Perceptrons, an earlier incarnation of neural nets, received a public trashing at the hands of Marvin Minsky and Seymour Papert in their book *Perceptrons,* published in 1969.

Edition by edition, it would seem, new and still newer computer omnibuses, a fleet of them, will converge on the curriculum guidelines of the Association for Computing Machinery. With tongue only halfway in cheek, I see it all ushering in a kinder and gentler age in the instruction and learning of computer science.

There are now 66 chapters, all mixed up as they were in the 1989 edition. The mixing is deliberate. A chapter that prepares in some important way for another always precedes it. But between the two, the sequential reader and browser will discover a number of other chapters on quite different topics. This leaves a not unfair impression of computer science as a whole, an interweaving of diverse ideas and approaches. Some chapters, moreover, refer to other chapters well outside their theme. Topics are cross-linked, as they are in the living subject. Ordinary textbooks cannot, by their confined nature, illustrate how ideas sometimes fertilize each other.

Each chapter deals with a specific topic. Some chapters, however, contain secondary topics that are sneaked in under the guise of examples or illustrations. Thus the chapter on correctness of algorithms (Chapter 10, Ultimate Debugging) illustrates a proof technique on Euclid's famous algorithm for finding the greatest common divisor of two integers, a miniature topic in its own right. Most chapters attribute a key idea to a specific person or persons. I am painfully aware, however, that other attributions of the same or related ideas may be missing (for the sake of simplicity). But each chapter sports its own two-book bibliography, supplementary reading for those who wish to know more about the topic covered. Here the names of other contributors to the subject will be found.

Each chapter is also followed by two or more exercises. Some are absurdly easy, some quite lengthy or difficult. I have not distinguished the one from the other because I want to encourage thought: the reader may at least consider the exercise long enough to wonder how it could be tackled. At the same time, any instructor who uses this book in a classroom or tutorial setting might well give some exercises as assignments.

Acknowledgments

I must thank first of all Dominic Welsh and Jon L. Bentley for primary encouragement and assistance with the book. As I state in the Preface, Welsh first suggested several of the topics. Several years later, Bentley, well known for his columns in the *Communications of the ACM* as well his numerous contributions to the analysis of algorithms, tested several chapters in courses he taught at the U. S. Air Force Academy. He also provided other topics and examples.

For the 1989 edition, Elizabeth Mergner and Barbara Friedman at Computer Science Press had the vision to encourage the preparation of a manuscript. They appointed series editor Jeffrey Ullman, himself a well-known computer scientist, to examine and comment on more than half the book. Later, Frank Boesch, a computer scientist and department chair at the Stevens Institute in Hoboken, New Jersey, undertook the most comprehensive review of all. To both these scholars I owe much. In the acknowledgments to the 1989 edition I stated at this point, "Whatever mistakes or roughness remain are mine alone." There were mistakes and roughnesses, but they have all been corrected in the new edition. Passengers on the omnibus should have no fear of minefields.

Numerous friends and colleagues offered help and suggestions that in some way advanced the book. In alphabetical order and with no indication of which edition they affected, they are Michael Bauer, Michael Bennett, Ernest Brickell, Larry Cummings, Charles Dunham, Ron Dutton, Edward Elcock, Irene Gargantini, Ron Graham, Bob Mercer, Jim Mullin, Sylvia Osborn, Vijay Ramanath, Stuart Rankin, Arto Salomaa, Hanan Samet, Gus Simmons, Andrew Szilard, Tim Walsh, and David Wiseman. I must also thank several of the more helpful readers of the 1989 edition who drew more than one mistake or flaw to my attention: Henry Casson, Mark E. Dreier, Kimball M. Rudeen, Richard D. Smith, S. R. Subramanya, and William R. Vilberg.

Special thanks must go to Stephen Wagley, editor of the 1989 edition, and to Penelope Hull, editor of the new edition, for their skills and energy, a tribute to Computer Science Press and W. H. Freeman and Company.

In the 1989 edition I wondered whether *The Turing Omnibus* would run again. Fueled in part by readers who send in ideas for topics or suggestions for changes, *The (New) Turing Omnibus* invites them to ride again, along with a host of new readers who may demand, sooner or later, an even larger vehicle.

A. K. Dewdney
June 1993

ICONS

At the beginning of each chapter, an icon identifies the area of computer theory or practice from which the chapter is drawn. These areas, with their icons and the chapters in each area, follow.

Analysis of Algorithms
1, 10, 22, 25, 30, 40, 50, 62

Applications
4, 9, 16, 21, 29, 32, 57

Artificial Intelligence
6, 19, 36, 58, 64

Coding and Cryptology
12, 37, 49

Complexity Theory
8, 15, 41, 45, 54, 66

Computer Graphics
18, 24, 47

Data Structures
11, 35, 43, 52, 61, 65

Automata and Languages
2, 7, 14, 23, 44

Logic and Systems Design
3, 13, 20, 28, 38, 48, 53, 56, 60

Theory of Computation
5, 17, 26, 31, 34, 39, 51, 59, 63

Miscellany
27, 33, 42, 46, 55

THE (NEW) TURING OMNIBUS

ALGORITHMS

Cooking Up Programs

A program specifies in the exact syntax of some programming language the computation one expects a computer to perform. The syntax is precise and unforgiving. The slightest error in the program as written may cause the computation to be in error or may halt it altogether. The reason for this situation seems paradoxical on the surface: It is relatively easy to design a system that converts rigid syntax to computations; it is much harder to design a system that tolerates mistakes or accepts a broader range of program descriptions.

An algorithm may specify essentially the same computation as a specific program written in a specific language such as BASIC, Pascal, or C. Yet, the purpose of an algorithm is to communicate a computation not to computers but to humans. This is a more natural state of affairs than most people suppose. Our lives (whether we work with computers or not) are full of algorithms. A recipe, for example, is an algorithm for preparing food (assuming, for the moment, that we think of cooking as a form of computation).

ENCHILADAS

1. Preheat oven to 350°F.
2. In a heavy saucepan
    1. Heat 2 tablespoons olive oil.
    2. Sauté
        ½ cup chopped onion
        1 minced garlic clove
        Until golden
    3. Add
        1 tablespoon chili powder
        1 cup tomato puree
        ½ cup chicken stock
    4. Season with
        Salt and pepper
        1 teaspoon cumin
3. Spread sauce over tortillas.
4. Fill centers with equal quantities of
    Chopped raw onion
    Chopped mozzarella cheese
5. Roll tortillas.
6. Place in ovenproof dish.
7. Pour more sauce over tops.
8. Sprinkle with chopped mozzarella cheese.
9. Heat thoroughly in oven about 15 minutes.

Figure 1.1 Programmed enchiladas

The purpose of this chapter is not only to acquaint the general reader with the idea of an algorithm, but also to prepare her or him for the manner in which this book presents algorithms. The recipe illustrates some of the conventions to be used as well as some of the broader features shared by most algorithmic styles.

The name of an algorithm is capitalized, as in ENCHILADAS. The individual steps or statements are usually (but not always) numbered. In this respect, one notices a peculiar feature of ENCHILADAS; some statements are indented and appear to repeat earlier numbers. For example, after the first statement labeled 2, there are four statements numbered 1 to 4, all sharing the same indentation. The purpose of the indentation is to make it easier for the eye to recognize that the four steps all take place in the heavy saucepan specified in step 2. If we wish to refer to one of these four steps, we use the step number preceded by the number of the last nonindented step before it. Thus the cook must add chopped onions and garlic in step 2.2. As soon as the saucepan operations have been completed at the end of step 2.4, the statement labels revert to single numbers.

This statement-numbering scheme is used by only a few authors, but it is convenient. Other schemes are possible. The main purpose in numbering statements is to make references to them possible. It is easier to draw a reader's attention to statement 2.3 than to talk about "the step where one adds the chili powder, tomato puree, and chicken stock."

The indentations used in the recipe follow a broader rationale than the indentations used in most algorithms — or programs, for that matter. But statement 2.2 is typical: There is a *continuing* operation (sautéing) that is repeated *until* a certain condition is met, namely, when the onions and garlic have turned golden. Such an operation is called a *loop*. All statements inside the loop are indented.

A major feature shared by ENCHILADAS with all algorithms is the looseness of its description. Glancing through the algorithm, one may notice several operations that are not spelled out. How "heavy" must the saucepan be? When is a color "golden"? How much salt and pepper should be added in step 2.4? These are things that an inexperienced cook (like the author) might worry about. Such things do not bother experienced cooks. Their judgment and common sense fill the obvious gaps in the recipe. An experienced cook, for example, knows enough to remove the enchiladas from the oven at the end of step 9, even though the algorithm does not explicitly include that direction.

Whatever the parallels between cooking and computation, algorithms come into their own in the latter environment. This does not mean that algorithms are incapable of producing the computational equivalent of a good enchilada. Consider, for example, the following graphic algorithm. It produces a mind-boggling variety of pleasing designs on a computer screen:

WALLPAPER

1. **input** *corna, cornb*
2. **input** *side*
3. **for** $i \leftarrow 1$ to 100
    1. **for** $j \leftarrow 1$ to 100
        $x \leftarrow corna + i \times side/100$
        $y \leftarrow cornb + j \times side/100$
        $c \leftarrow int\,(x^2 + y^2)$
        if c *even*
            then *plot* (i, j)

This algorithm, when it is translated to a program for a computer with some kind of graphic output device (such as a simple display screen), will enable users to explore a world of endless wallpaper patterns. Each square section of the plane has its own pattern. If the user types in the coordinates of the lower left-hand corner *(corna, cornb)* of the square of interest, along with the length of its sides, the algorithm draws a picture of wallpaper associated with that square. Strangely enough, if a small square with the same corner coordinates is chosen, a completely different pattern emerges, neither a magnification nor any other transmogrification.

Figure 1.2 Algorithmic wallpaper

So much for what the algorithm does. How does it work? There are two **for** loops in the algorithm. The outer loop steadily counts from 1 to 100 in step 3, using variable i to hold the current count value. For *each* such value, the inner loop counts from 1 to 100 via j in step 3.1. Moreover, for *each* j value, the four steps that we might label 3.1.1 through 3.1.4 are repeated many times. Each time, of course, there is a new pair of values i and j to work with. First x is computed as a coordinate that is i one-hundredths of the way *across* the square. Then y is similarly computed as a coordinate that is j one-hundredths of the way *up* the square. The resulting point (x, y) is therefore somewhere within the square. As j goes from value to value, one might think of the point moving steadily up the picture. When it reaches the top, $j = 100$ and the inner loop has been fully executed. At this point, i moves to its next value, and the whole inner loop begins all over again at $j = 1$.

The heart of the algorithm consists of computing $x^2 + y^2$, a circular function, and then taking the integer part (int) of the result. If the integer is even, a point (say, black) is plotted on the screen. If the integer is odd, however, no point is plotted for (x, y) — the screen remains light there.

The WALLPAPER algorithm illustrates several practices followed in this book.

Statements are indented inside the scope of loops and of conditional statements (**if . . . then**).

Assignment of a computed value to a variable is indicated by a left-pointing arrow.

The computation indicated on the right-hand side of an assignment statement can be any meaningful arithmetic formula.

More than this, we make the statements of an algorithm very general. For example, WALLPAPER might be rewritten as follows:

1. **input** parameters of square
2. **for each** point in the square
   1. compute the c function
   2. **if** c is even, plot it

Such compressed algorithms might serve several purposes. They may represent a stage in the design of a WALLPAPER program in which the programmer begins with a very general, coarse-grained description of the computation to be carried out. He or she then rewrites the algorithm a number of times, refining it at each stage. At some point when WALLPAPER has reached a reasonable level of detail, it can be translated more or less directly to a program in any of a number of actual programming languages, from BASIC to C.

A second reason for writing compressed algorithms lies in the level of communication with another human being. For example, the writer of an algorithm may share a certain understanding with another person about the intention behind each statement. To say "**for each** point in the square" implies a double loop. The "*c* function" might well be understood to mean steps 3.1.1 to 3.1.4 in the WALLPAPER algorithm.

The kind of algorithmic language used in this book and elsewhere freely borrows constructs that are available in a great many programming languages. The following loop types are useful:

for . . . **to**
repeat . . . **until**
while . . .
for each . . .

There are **input** and **output** statements, conditionals such as **if** . . . **then** . . . or **if** . . . **then** . . . **else**. . . . Indeed, readers should feel free to extend the algorithmic language used here to include any language construct that seems reasonable. Variables can be real numbers, logic-valued (boolean) numbers, integers, or virtually any single-valued mathematical entity. One may have arrays of any type or dimension, lists, queues, stacks, and so on. Readers unfamiliar with some of these notions may discover their meanings in the context of whatever chapters use them.

The translation of an algorithm to a program is generally straightforward, at least if the algorithm is reasonably detailed. For example, the WALLPAPER algorithm translates readily to the following Pascal program:

```
program WALLPAPER (input,output);
var corna, cornb, side, x, y, c: real;
var i, j: integer;
begin
    read(corna,cornb);
    read(side);
    graphmode;
    for i: = 1 to 100 do
    begin
        for j: = 1 to 100 do
        begin
            x: = corna + i* side/100;
            y: = cornb + j* side/100;
            c: = trunc(x*x + y*y);
```

```
        if cmod2 = 0 then graph.putpixel (i,j,1);
        end
    end
    textmode;
end
```

The instructions *graphmode* and *textmode* must be translated by the reader into the appropriate graphics initialization mode proper to his or her own version of Pascal. The *textmode,* optional, simply restores the text screen if the user should need it.

Insofar as an algorithm represents computation, one might say that *The Turing Omnibus* is almost a book about algorithms. What problems can (or cannot) be solved by algorithms? When is an algorithm correct? How much time does it take? How much memory does it use? Many algorithms, moreover, embody subtle and interesting intellectual questions. The WALLPAPER algorithm is no exception. Why is it, for example, that in so many cases (as in the previous figure) the algorithm produces repeating or nearly repeating patterns?

Every chapter in this book is followed by a small set of problems. Although a great many chapters describe certain algorithms, very few problems suggest that the reader implement the algorithms as actual programs. But the suggestion is always implicit, and students of computing are urged to do precisely this — starting with this chapter.

Problems

1. Write your favorite recipe as an algorithm. How much more complicated does the algorithm become if it is to be used by a novice cook?

2. Rewrite the WALLPAPER algorithm to use not two colors (black and white) but three.

(*Hint:* Even and odd are merely the residue classes of a number taken to modulo 2.)

References

Irma S. Rombauer and Marion Rombauer Becker. *Joy of Cooking.* New American Library, New York, 1964.

David Harel. *Algorithmics: The Spirit of Computing.* Addison-Wesley, Reading, Mass., 1987.

FINITE AUTOMATA

The Black Box

I t occasionally happens in industrial, military, or educational settings that one is presented with a piece of electronic hardware whose exact function is uncertain or unknown. One way of discovering how the device works is to take it apart, piece by piece, and to deduce its function by analyzing the components and their interconnections. This is not always possible, however, nor is it always necessary. Given that the mystery machine has both input and output facilities, it may be possible to discover what it does without ever taking it apart. Since its appearance gives no clue about its function, we call it a *black box.*

Let us imagine a particular black box (Figure 2.1) bearing an electrical terminal marked *input,* a small lamp labeled *ACCEPT,* and a pushbutton marked *reset.* The machine appears to accept two kinds of voltage, and representing these by the symbols 0 and 1, we carry out some experiments with the black box. The first few symbols have no effect whatever on the machine. After several more however, the lamp suddenly lights up; our sequence has been ACCEPTed! As careful experimenters, we have recorded the sequence of signals which turned on the lamp, but repeating the sequence fails to turn it on. For a time we

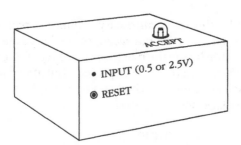

Figure 2.1 A mysterious black box

are stymied and even begin to imagine that the circuitry inside the box is behaving randomly. Then someone tries the same sequence after pushing the reset button. The lamp goes on again. Continuing these experiments, we discover other sequences which, following a reset, also light the lamp. We note these and begin to wonder how far we can go in analyzing the behavior of the black box.

To do such an analysis, we must assume that the contents of the box are in some kind of stable equilibrium between one input signal and the next. We also assume that the interior circuitry can exist in only a finite number of such states of equilibrium. But these assumptions are perfectly reasonable — at least if the device is manufactured.

Indeed, when it is described this way, the black box is essentially a *finite automaton*. In abstract terms, such a device consists of

A finite collection Q of states (i.e., stable internal electronic conditions)
A finite alphabet Σ of input signals
A function δ which, for every possible combination of current state and input, determines a new state

Two of the states in Q are special: One is called an *initial state,* and another is an *accepting state* (sometimes called a *final state*).

So the reset button on our black box puts the automaton into the initial state. Certain sequences of 0s and 1s result in the automaton entering the accepting state, causing the light to go on.

A finite automaton is said to *accept* any sequence of symbols which places it in its final state. The set of all such sequences may be regarded as words in the *language* of that automaton. With this terminology, then, we may ask, Is it possible to determine the language of the automaton inside the black box? Can

9

we succinctly describe the set of all binary sequences which cause the little lamp to light? These, after all, are the *words* of the language.

Because the automaton inside the black box has only a finite number of states, we might suppose this to be a reasonable project. But after a number of experiments, we find ourselves beset with doubts. For example, the following words are accepted by the black box:

> 0101
> 0100101
> 0100100101
> 0100100100101

It does not take long to surmise that the automaton is accepting all words of the form 01(001)*01, where (001)* is just shorthand for "any number of repetitions of the sequence 001." In fact, we may even be tempted to draw part of the automaton's *state-transition diagram* (Figure 2.2) in which states are represented by labeled circles and transitions between them by arrows.

The transitions represent the function δ. For example, if the automaton is currently in state 3, then δ determines that the next state is 4 if a 1 is inputted and 5 if a 0 is inputted.

But do we *know* that this diagram is correct? Unfortunately, we do not. It is easy to imagine other diagrams which give exactly the same result, including the one shown in Figure 2.3.

Indeed, we may even have been wrong in our surmise that the black box would accept all strings of the form 01(001)*01. It might accept all such strings with up to 98 repetitions of 001 but not accept any with 99 repetitions. "Black-boxing" an unknown device can be a tricky business!

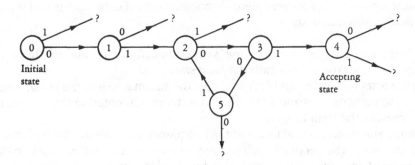

Figure 2.2 A possible state-transition diagram

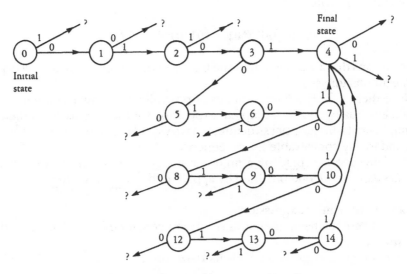

Figure 2.3 Another possible state-transition diagram

At this point a thoughtful member of our investigative team might happen to turn over the black box and discover the following label:

GENIAL ELECTRIC
Black Box 3718-45X
(6 states)

Given this information, it is at last possible to determine precisely the language accepted by the automaton within the box. A certain theory (see Chapter 14) tells us that if the automaton accepts the six words $01(001)^n 01$ (namely 01 followed by n repetitions of 001 followed by 01), for $n = 0,1,2,3,4,5$, then it will accept all words in the set $01(001)^*01$. Thus we may write

$$01(001)^*01 \subseteq L_A$$

where L_A represents the language of the automaton A, the one inside the black box.

A systematic exploration of the possibilities leads, let us say, to another such conclusion:

$$(0110)^*111 \subseteq L_A$$

We could then write

$$01(001)*01 + (0110)*111 \subseteq L_A$$

which means that a word in either the first form or the second will be accepted by the automaton A.

Since there is only a finite number of possibilities for such sequences, we would eventually obtain a complete expression for the automaton in the box; we might not know its exact structure, but we would know exactly how it would respond to any conceivable input sequence.

The language accepted by a finite automaton can always be written as a *regular expression*, one obtained by applying the following rules:

$\left\{ \begin{array}{l} \text{0 and 1 are regular expressions.} \\ \text{If } X \text{ and } Y \text{ are regular expressions, so is their } \textit{concatenation } XY \text{ and so is their} \\ \quad \textit{sum } X + Y. \\ \text{If } X \text{ is a regular expression, so is its } \textit{iteration } X^*. \end{array} \right.$

Moreover, for every regular expression there is an automaton which accepts the language symbolized by that expression. Thus, in a sense, regular expressions capture precisely the behavior of automata in terms of the language they accept.

However, for every regular expression there are an infinite number of automata which accept that language. So even when we know precisely the language accepted by an automaton, we have to crack open the box and inspect its state-transition diagram, so to speak, if we wish to know exactly which device inhabits it.

Finite automata are the simplest kinds of computational model widely studied. As such, they have limitations which may not be readily apparent. For example, no finite automaton can accept the set of all "marked palindromes" over a given alphabet. These are words of the form zmz^{-1}, where z is a word over some alphabet, z^{-1} is its reverse, and m is a special marker symbol not in the alphabet. In Chapter 14, an even simpler example of a nonregular language is given. More powerful computational devices are discussed in Chapter 7.

Problems

1. Describe finite automata that accept languages

   a. $(ab)*(a+b)*$

   b. $0 + 1$

   c. $0* + 1$ (a finite automaton may have more than one accepting state)

2. It will be evident in Chapter 14 that regular languages are precisely those accepted by finite automata. Given this fact, show that if the language L is accepted by some finite automaton, then L^{-1} is also accepted by some finite automaton; L^{-1} consists of all words of L reversed.

3. How would you argue that no finite automaton can accept all palindromes over a given alphabet?

References

J. E. Hopcroft and J. D. Ullman. *Introduction to Automata Theory, Languages, and Computation.* Addison-Wesley, Reading, Mass., 1979.

Marvin Minsky, *Computation: Finite and Infinite Machines.* Prentice-Hall, Englewood Cliffs, N.J., 1967.

No finite automaton can accept the set of all "marked palindromes" over a given alphabet.

SYSTEMS OF LOGIC

Boolean Bases

In an age of computers and automation, almost every electronic device one can name incorporates at least one boolean function. For example, many current models of automobile will emit a high-pitched whine, buzz, or other disturbing noise until their drivers fasten their seat belts. Such a device realizes a boolean function of two variables A (= 1 if the ignition key is turned on) and B (= 1 if the seat belt has been fastened). Both variables take the value 0 or 1, usually interpreted as false or true, respectively, and the function itself takes on the same values. Suppose the function is W (to *w*arn the driver) so that when $W = 1$, the signal is to be on while if $W = 0$, it will be silent. The operation of the entire warning system can thus be summarized by W, where

$$W = \begin{cases} 1 & \text{if } A \text{ is 1 and } B \text{ is 0} \\ 0 & \text{otherwise} \end{cases}$$

In fact, we can do better than this by defining two logic operators AND and NOT and using them as follows:

$$W = A \text{ AND} (\text{NOT } B)$$

where for two boolean variables X and Y we define X AND Y to be 1 whenever X and Y are both 1; NOT X is 1 whenever X is 0 and is 0 whenever X is 1.

We have just defined one boolean function, W, in terms of two others, AND and NOT. Usually the latter two are written with a multiplication symbol ($\cdot$) and a complementation symbol ($'$). Thus we can now write the function W even more compactly as $W = A \cdot B'$.

Any boolean function can be defined by means of a truth table, which is merely a convenient way of listing the values of the function for each possible combination of values of its variables. For example, W could be defined by the following table.

S	B	W
0	0	0
0	1	0
1	0	1
1	1	0

It might seem that if we were given an arbitrary boolean function of two variables, we would have to be lucky to be able to express it by using AND and NOT. Interestingly enough, this is not the case: *Any* two-variable boolean function can be written in this manner. The easiest way to understand this is to consider the following example involving a function F in this truth-table form:

X	Y	F
0	0	0
0	1	1
1	0	0
1	1	1

The essential idea is to write F as a product of expressions, one per row of the table, which individually produce a value of 0 for F. Thus, when any of the individual expressions equals 0, so does F. An example is furnished by the very first row of F. Because the expression can use only the $\cdot$ and $'$ operators, we must ask what combination of them will produce a value of 0 when X and Y are both 0?

The answer is $(X' \cdot Y')'$. Only for the combination $X = 0$ and $Y = 0$, does this expression have the value 0:

$$(0' \cdot 0')' = (1 \cdot 1)'$$
$$= 1'$$
$$= 0$$

Similarly, the third row of F's table also corresponds to an F value of 0. Its expression is $(X \cdot Y')'$. When (and only when) $X = 1$ and $Y = 0$, this expression equals 0.

We may now write

$$F = (X' \cdot Y')' \cdot (X \cdot Y')'$$

Both the general rule and the reason why it always works are becoming clear. Thus, if we had defined F above to have an additional zero, say, in the second row, then we could write

$$F = (X' \cdot Y')' \cdot (X' \cdot Y)' \cdot (X \cdot Y')'$$

Since any two-variable boolean function can be expressed thus in terms of ($\cdot$) and ($'$), we say that these two operations form a *complete base* for the two-vari-

	Function name	Symbol	In terms of $\cdot$, $+$, and $'$
1	OR	$+$	$x + y$
2	AND	$\cdot$	$x \cdot y$
3	NOT (x)	$'$	x'
4	NOT (y)	$'$	y'
5	EXCLUSIVE OR	$\oplus$	$x' \cdot y + x \cdot y'$
6	EQUIVALENCE	$\equiv$	$x \cdot y + x' \cdot y'$
7	IMPLICATION	$\rightarrow$	$x' + y$
8	NONIMPLICATION	$\not\rightarrow$	$x \cdot y'$
9	REVERSE IMPLICATION	$\leftarrow$	$x + y'$
10	REVERSE NONIMPLICATION	$\not\leftarrow$	$x' \cdot y$
11	PROJECTION (x)	x	x
12	PROJECTION (y)	y	y
13	NAND	$\mid$	$(x \cdot y)'$
14	NOR	none	$(x + y)'$
15	CONSTANT 0	0	0
16	CONSTANT 1	1	1

[Handwritten top margin: (·), (') = complete base for the 2 var. functions ≡ any 2 var. boolean function can be expressed in terms of (·) and (')]

able functions. Moreover, $\{\cdot, '\}$ forms a *minimal complete base* because neither operation alone will suffice to express all two-variable boolean functions. What has just been said about two-variable functions also holds for three-, four-, . . . , n-variable functions: The set $\{\cdot, '\}$ is a complete base for a boolean function on *any* number of variables. If a function F has n variables, look at each row of its truth table where it has the value 0 and write the "product" of variables (appropriately complemented) corresponding to that row. If the row is 010110, write $x_1' \cdot x_2 \cdot x_3' \cdot x_4 \cdot x_5 \cdot x_6'$. Now enclose each of the expressions just formed in parentheses, apply the complementation operator to each, and string all the resulting expressions together into one big product. This product has value 1 if and only if F does, for the same reasons indicated above.

The question now arises: Which other sets of operators form complete bases? Assuming that by "operators" we mean two-variable boolean functions, we may answer this question only after we have examined all the possible two-variable functions, as shown in the table on page 16.

It should not be too surprising that there are exactly 16 two-variable (x, y) boolean functions; these merely correspond to all possible ways of filling in a truth table for an unknown function F:

x	y	F
0	0	?
0	1	?
1	0	?
1	1	?

Certain combinations of these 16 boolean functions form complete bases in the following sense: When the functions are considered binary operations, any boolean function on any number of variables can be expressed in terms of those functions. For example, the set $\{+, '\}$ and the set $\{+, \equiv, \oplus\}$ are both complete bases. In fact, they are minimal complete bases. No proper subset of them is a complete base.

In Figure 3.1, all the minimal complete bases are summarized. Each point represents one of the two-variable functions in the list above, and the minimal complete bases are represented by points, lines, or solid triangles. A point not contained in a line and a line not contained in a solid triangle both represent complete bases with one and two elements, respectively. Of course, the solid triangles represent minimal complete bases with three elements.

Since only 11 operations are represented in this figure, 5 are clearly missing from the 16 operations represented earlier. When they are considered as *opera-*

[Handwritten bottom margin: $010110 = (x_1')'(x_2)\cdot(x_3')'\cdot(x_4)'(x_5' \cdot (x_6')' = x_1 \cdot x_2' \cdot x_3 \cdot x_4' x_5'. x_6 = 1$ iff $F = 1$]

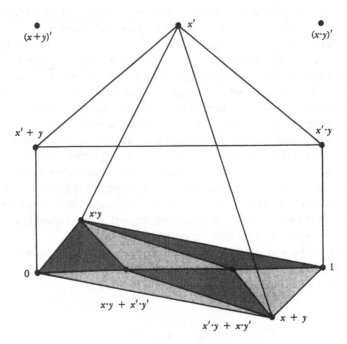

Figure 3.1 Structure of the complete bases

tions, however, three of the two-variable functions on our list turn out to be redundant. For example, when it is considered as an operation, x' is the same as y'. Moreover, $x' + y$ is the same as $x + y'$, and $x'y$ is the same as xy'. These operations will themselves be carried out on pairs of boolean functions, and $x' + y$ just means "complement one of the functions and add it to the other"; either function can be x, and the other can be y. In addition to these redundancies, the projection functions give us nothing as operations; they tell us, in effect, to do nothing.

In the context of computer logic, these complete bases acquire a significance which goes beyond the purely mathematical interest of the above scheme. For the circuitry in the control unit, arithmetic unit, and other logic components of a computer to function appropriately, the logic must involve a set of operations which forms a complete base.

In most texts in computer logic design, the various logical operations are represented by *gates.* These are simply boolean operations of one of the above specified types applied to specific sets of electric lines in the circuit. For example, a gate diagram for the seat-belt warning circuit mentioned at the beginning of this chapter would appear as shown in Figure 3.2.

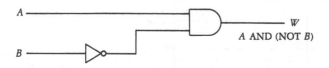

Figure 3.2   Seat-belt warning logic

The semicircular gate represents AND and the triangular gate, sometimes called an "inverter," represents NOT. The OR function is represented by a shield-shaped figure like the one in Figure 3.3, which shows a somewhat more sophisticated circuit called a *multiplexer.*

Multiplexers are found in all computers. They enable several different logic functions (entering on lines labeled 0, 1, 2, and 3) to use a single line (labeled *output*). This is done sometimes for reasons of economy and sometimes because the functions being multiplexed are to be used by a single device at the other end of the output line. In any event, the multiplexer shown here operates as follows: Depending on which pair of voltages (0 or 1) enters the select lines labeled S_0 and S_1,

S_0	S_1	*Select line no.*
0	0	0
1	0	1
0	1	2
1	1	3

only one of the four AND gates will transmit the signal arriving along its input line from the left. For example, if $S_0 = 0$ and $S_1 = 1$, then AND gates 0, 1, and 3 all receive at least one 0 from the S lines and hence cannot contribute anything but a 0 to the OR gate operating the output line. However, AND gate 2 receives two 1s from its S lines, and its output, therefore, will be 0 or 1, depending on exactly what signal arrives on line 1. Notice that the bits (0 and 1) appearing on the select lines simply give the binary expansion of the number of line being selected for output.

Given this brief introduction to logic diagrams involving AND, OR, and NOT gates, we may now examine some of the minimal complete bases displayed earlier in terms of computer logic.

One of the more interesting portions of the bases diagram is the isolated point labeled $(x \cdot y)'$. This operation, called *NAND,* forms a complete base all by

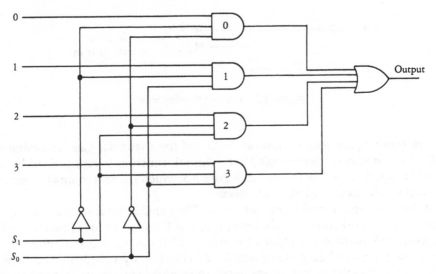

Figure 3.3 A multiplexer

itself! The simplest way to see this is to write the operations in one complete base, say {·,′}, in terms of the NAND operation. Just for fun, we will do this in terms of gates (see Figure 3.4).

Here, we have used NAND gates (a semicircle with a knob) to realize the AND and NOT operations. A single-input NAND gate operates exactly like an inverter, and inverting the output of a NAND gate yields the AND operator.

As luck would have it, NAND gates are very simply constructed from transistors, whether in the old-fashioned semiconductor technology or in the newer silicon-chip fabrication techniques. Most logic diagrams that are used to represent what is actually going on in today's computers contain large numbers of NAND gates.

Problems

1. Show that NOR is also a complete base, all by itself.

2. Show that {·, ≡, ⊕} is a complete base.

3. Redraw the multiplexer circuit, using only NAND gates, and make it as simple as possible. Is this the simplest NAND circuit which acts as a multiplexer on four input lines?

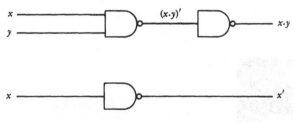

Figure 3.4   Converting the base $\{\cdot, '\}$ to $\{|\}$

References

Richard Johnsonbaugh. *Discrete Mathematics,* 2d ed. Macmillan, New York, 1990.

M. M. Mano. *Digital Logic and Computer Design.* Prentice-Hall, Englewood Cliffs, N.J., 1979.

SIMULATION

The Monte Carlo Method

I n the quest to understand the many systems that comprise the modern world, we turn increasingly to computer simulation. Whether the system is natural or artificial, frequently one or more of its components have such complex behavior that the only feasible approach to approximating such behavior is to assume that it is random. The obvious technique is to replace such complex components with simple ones following statistical laws. For example, if we were attempting to simulate a volleyball match between two evenly balanced teams, we might make a very simple model of the game by continually flipping a coin: If one team is "serving" and the toss goes against them, then the other team gets the serve; otherwise, that team is awarded a point. The coin-flipping operation, reminding us of a simple betting procedure, could be called a *Monte Carlo technique*.

In this chapter we use a simple bank system to illustrate the major features of Monte Carlo simulation. Consider, for example, a bank with one teller (Figure 4.1). Customers enter, seemingly at random, and form a queue in front of the teller. Below the illustration is shown a schematic version of the scene as an

Figure 4.1 Customers form a queue at a bank

indication of how various real-world systems are made abstract. In a well-designed simulation, a tremendous amount of the detail inherent in the original system is ignored with little or no penalty; the behavior of the abstract model (when it is implemented on a computer) is indistinguishable from the real thing as far as the parameters under investigation are concerned.

Of course, every simulation, whether scientifically or industrially inspired, must have a goal, and the goal determines what aspects of the system being modeled are relevant. In the case of our simple bank simulation, it might be simply to decide whether it would be "worthwhile" to add another teller.

Theoretical considerations suggest and empirical studies imply that a real-world service facility, while in the steady state, tends to receive customers with a particular arrival pattern. Given an average time of α second(s) between arrivals, the actual distribution of such times follows the negative exponential distribution, as shown in Figure 4.2.

As with any density function, one must be careful with interpretation. For each possible time t between consecutive arrivals (interarrival time), the function f gives what might be called an *interarrival density*, but *not* the interarrival frequency as such. One can only take some range of interarrival times, say from t to t', and calculate the area under the curve. This yields the relative number of times one can expect an interarrival time in the range t to t'. Thus, if one wanted to predict, for 100 consecutive customers, how often an interarrival time in this range occurred, one would simply calculate the shaded area A shown in

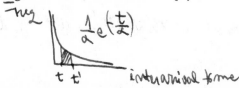

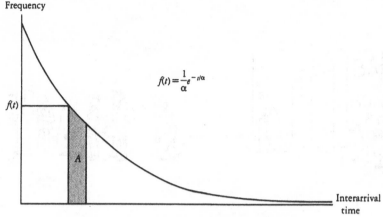

Figure 4.2 The negative exponential distribution

the diagram and then compute $A \times 99$, where 99 is obviously the number of interarrivals.

But the foregoing considerations do not show us how to generate arrivals in a computer. For this we need the cumulative distribution function F, which is just the integral of f (Figure 4.3).

Here, a given interarrival time t yields a cumulative frequency $F(t)$, namely, the proportion of interarrivals having times t or less. One can make a most interesting observation about the function F: If one selects random numbers x

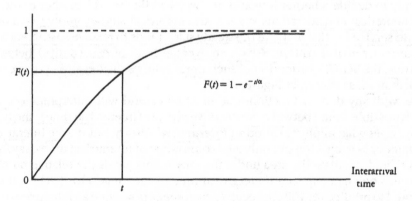

Figure 4.3 Cumulative form of the negative exponential distribution

between 0 and 1 and then calculates $F^{-1}(x)$, the resulting quantities reproduce the original density function f! In practical terms, this means that if one selected, say, 1000 random numbers in the range 0 to 1 and converted each of them via the inverse function F^{-1}, then a histogram of the resulting quantities would tend to have the same shape as the density curve f shown in Figure 4.2.

This observation enables us to generate random arrivals at the simulated bank by following steps 2 to 6 over and over:

1. Generate the first customer.
2. Select a random number x between 0 and 1.
3. Compute $F^{-1}(x)$.
4. Allow $F^{-1}(x)$ seconds to elapse.
5. Generate the next customer.
6. Go to step 2.

Random numbers can be generated by any of the techniques suggested elsewhere in this book (see Chapter 8). For the function F discussed here, it is easy to show that

$$F^{-1}(x) = \alpha \ln (1 - x)$$

One of the advantages of computer simulation is that experiments that might take days or even months of real time on actual systems (assuming one is allowed to tinker with them in the first place) can be carried out by machine in seconds, minutes, or (in the worst case) hours of real time. This is because the flow of time is also simulated—in most cases by a simulation clock, a real variable (say, C) which keeps track of time. Thus, at the beginning of our simulation, at step 1, we have $C = 0$. If the first interarrival time generated at step 3 is 137 seconds, then in step 4 we would add 137 to C, updating the simulation clock.

Step 5, that of generating the next customer, means simply that we add a token representing that customer to some data structure representing the queue in front of the teller. As it turns out, a simulation this simple does not require us to represent the queue by anything as elaborate as an array or list structure. It is sufficient that Q merely records how many customers "currently" occupy the queue. Thus, when the computer encounters stage 5 above, it merely adds 1 to Q.

Customers are subtracted from the queue as soon as they have received a completed service from the teller. Although the service times one actually finds in banks do not have a simple distribution, we assume that these also follow a certain exponential distribution G, in this case with the time constant β, the teller's average service time per customer.

We are now in a position to write the same sort of program for service times as

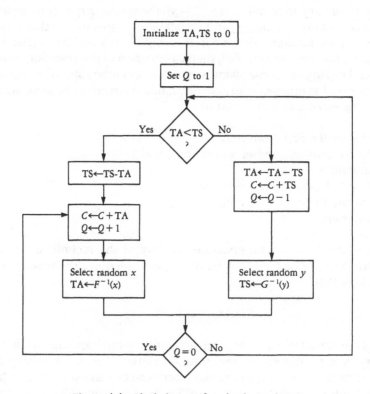

Figure 4.4 Block diagram for a bank simulation

indicated earlier for interarrival times. Now, however, two kinds of operations must be successfully interleaved so that each step follows in logical sequence as outlined in Figure 4.4. An essential feature of the approach taken here is to compute the time to the next event (arrival of customer or completion of service) through the use of two time variables TA and TS.

In Figure 4.4, notice how both TA (time to next arrival) and TS (time to end of current service) act as little clocks. If TA is less than TS, then an arrival will happen first and we must increment the clock by the amount TA and then *decrement* the TS clock by the same amount. After adding the new arrival to the queue, we generate the next interarrival time and reset TA accordingly. On the other side of the diagram, the roles of TS and TA are naturally reversed.

This approach is called the *critical-event technique*. One simply increments the simulation clock to the time of the next change in the system and then carries out the change indicated—in this case altering the queue length. Another approach to managing time in computer simulations is called the *time-*

slice method. It involves choosing some small fundamental increment in time, continuing to update the clock by this increment, and sweeping through the system, computing its new state on the basis of its present state.

The heart of the Monte Carlo method is the generation of new events by means of the inverse-function technique. Virtually any distribution, whether theoretical in the form of a computable function or empirical in the form of a cumulative histogram, can be used. With a somewhat more complicated program structure than the one indicated above, any number of tellers can be modeled. In fact, one can simulate a multiplicity of queues (such as banks used to have) or a large, single queue feeding a multiplicity of tellers (such as banks now have). Indeed, in the most modern banks, the tellers themselves form a queue for service from a "central teller." This system, too, can be simulated by an extension of the techniques outlined above.

There are special-purpose languages such as GPSS which are tailored to systems with multiple queues and various other kinds of structure. It is frequently much easier to write simulation programs in these languages although almost as frequently the running time tends to be longer than for a custom-designed program in FORTRAN or Pascal.

Problems

1. Prove the observation, remarked on earlier, that interarrival times selected by the inverse (cumulative) function method follow the corresponding density distribution.

2. Given the block diagram for the single-teller system described in this chapter (Figure 4.4), how would you incorporate the following statistics-gathering requirements into the diagram?

   a. Measure average and maximum customer waiting times.

   b. Measure average and maximum queue lengths.

   c. Measure total idle time for the teller.

3. Suppose a bank manager wishes to compare the operation of his or her bank with and without the single-queue discipline. Devise a program design to simulate both systems and to decide which is better as far as two operational parameters are concerned: average customer waiting time and maximum customer waiting time.

4. Queuing theory tells us that in a single queue of the type studied here, the average queue length is $\rho/(1 - \rho)$, where $\rho = \beta/\alpha$. Write a simulation program

? Queuing theory ?

for specific α and β, with $\alpha > \beta$. Allowing the program to run for a while to achieve steady state, make a number of observations of queue length and compute their average. How close to the theoretical value does your observed value come?

References

F. S. Hillier and G. J. Lieberman. *Operations Research,* 2d ed. Holden-Day, San Francisco, 1974.

Herbert Maisel and Juliano Gnugnol. *Simulation of Discrete Stochastic Systems.* Science Research Associates, Chicago, 1972.

GÖDEL'S THEOREM

Limits on Logic

In the early 1930s, Kurt Gödel, a German mathematician, attempted to show that the predicate calculus (see Chapter 58) was "complete" — that one can obtain mechanically (in principle, at least) a proof of any true formula expressed in that calculus. His failure to do this was crowned by the discovery that the task was impossible: Certain formal systems, including arithmetic, were incomplete in this sense. The discovery staggered the world of mathematics.

As part of David Hilbert's turn-of-the-century program for mathematics, it had been expected that all mathematics, when it was suitably formalized in a system like the predicate calculus, would turn out to be complete. But Gödel discovered that not even arithmetic was complete. His now-famous theorem states that in any sound, consistent, formal system containing arithmetic there are true statements that cannot be proved — statements the truth of which we may know by other means but not by any formal, step-by-step decision process.

From Hilbert's time onward, a succession of researchers had attempted to formulate such decision processes in various ways. They had in common, however, the adoption of certain axioms and a variety of formalisms for manipulating the axioms by certain rules to obtain new statements that, given the truth of the axioms, were true themselves. One such system was the theory of recursive functions, which Gödel helped to develop. Recursive functions amount to yet one more description of what it means to compute. (See Chapter 66.)

The heart of Gödel's method was to encode each possible statement in the predicate calculus, regarded as a language, by a special numerical code called its *Gödel number*. Briefly, the process has three steps.

> Set up axioms for the predicate calculus along with a rule of inference by which one can get new formulas from old ones.
>
> Set up axioms for standard arithmetic in the language of predicate calculus.
>
> Define a numbering for each formula or sequence of formulas in the resulting formal system.

By using easy-to-read implicative language, the axioms for the predicate calculus may be listed as follows:

1. $\forall y_i(F \to (G \to F))$ *a true formula is implied by any formula*
2. $\forall y_i((F \to (G \to H)) \to ((F \to G) \to (F \to H)))$ *implication distributes over itself*
3. $\forall y_i((\neg F \to \neg G) \to ((\neg F \to G) \to F)))$ *if both $\neg G$ and G are implied by $\neg F$ then $\neg F$ can not be True!*
4. $\forall y_i(\forall x(F \to G) \to (F \to \forall xG))$ provided F has no free occurrence of x
5. $\forall y_i((F \to G) \to (\forall y_i F \to \forall y_i G))$
6. $\forall y_i(\forall xF(x) \to F(y))$ provided that y is not quantified when it is substituted

Once the notation used above is understood, the axioms will seem reasonably self-evident, the very thing upon which to base a theory of mathematical deduction.

Symbols such as F, G, and H are understood to stand for "formulas"; $\forall y_i$ stands for an arbitrary string of variables such as y_1, y_2, . . . , y_k, all universally quantified; one reads such symbolism as "for all y_i" or "for all y_1, y_2, . . . , y_k." The symbol $\to$ stands for implication and $\neg$ stands for negation.

The various formulas involved in the axioms above may or may not contain the variables which are quantified outside them, but in any case axiom 4 cannot be applied unless whenever variable x occurs in formula F, it must be quantified within F. Axiom 6 cannot be applied unless y is not quantified within $F(y)$ (the formula one gets by replacing all free occurrences of x in F by y).

The axioms themselves are now easily understood. For example, axiom 1 may be read, "For all possible values of their free variables, if F is true, then $G \to F$." In other words, a true formula is implied by *any* formula. Axiom 2 says that implication distributes over itself, in effect. And axiom 3 says that if both $\neg G$ and G are implied by $\neg F$, then $\neg F$ cannot be true, in other words, F must be true.

As a final example, axiom 4 says, "For all possible values of their free variables (as always), if $F \rightarrow G$ for all x and if x has no free occurrence in F, then $F \rightarrow \forall x G$."

To the previous set of axioms must be added a rule of inference like the following:

$$\text{If } F \text{ and } F \rightarrow G, \text{ then } G.$$

One notes that this rule does not lie at the same level as the axioms, in a sense: It is intended that whenever we are making a deduction—essentially a chain of formulas—and notice formulas F and $F \rightarrow G$ both occurring as earlier members of the chain, we may add G to the chain.

By the "standard arithmetic" is meant simply the *Peano postulates* for the natural numbers, namely,

$s = successor$

1. $\forall x \ \exists(0 = sx)$
2. $\forall x,y \ (sx = sy) \rightarrow (x = y)$
3. $\forall x \ x + 0 = x$
4. $\forall x,y \ x + sy = s(x + y)$
5. $\forall x,y \ x \times sy = x \times y + x$
6. $\forall x \ x \times 0 = 0$

Here, s denotes the successor function which for each natural number x picks out its successor, $x + 1$. Thus, postulates 1 and 2 tell us, respectively, that

Zero is not the successor of any natural number.
If the successors are equal, then so are the numbers.

The remaining postulates are similarly easy to understand. However, all six postulates beg the question of what we mean by equality. This very meaning is embedded in three additional postulates:

7. $\forall x \ x = x$
8. $\forall x,y,z \ (x = y) \rightarrow ((x = z) \rightarrow (y = z))$
9. $\forall x,y \ (x = y) \rightarrow (A(x,x) \rightarrow A(x,y))$

where A is any formula having two free variables.

Just as we added a special rule of inference to the axioms for the predicate calculus, so here we add a rule of induction:

$$(P(0) \& \forall x(P(x) \rightarrow P(sx))) \rightarrow \forall x P(x)$$

This formula simply encodes the well-known rule of induction: If a predicate P is true with 0 substituted in it *and* if whenever P is true of a number x, P is also true of x's successor, then P is true for all possible numbers x.

The 15 axioms and two rules listed above are, among them, powerful enough to give us a formal system for arithmetic in which so many ideas are expressible and truths provable that we are tempted a priori to imagine that any arithmetic truth is not only expressible in this system, but also provable there.

Having set up these axioms, Gödel went on to the third step in his proof, namely, to assign a unique number to every conceivable formula in the system just defined. He did this by assigning a natural number to each of the following basic symbols:

Symbol	Code Number	Symbol	Code Number
0	1	x	9
s	2	1	10
+	3	⌐	11
×	4	&	12
=	5	∃	13
(	6	∀	14
)	7	→	15
,	8		

Now, given any axiom or formula within our formal system, it is a straightforward matter to scan the formula from left to right and to replace each symbol in it by a prime number raised to the power of that symbol's code number. The primes used for this purpose are the consecutive primes 2, 3, 5, 7, 11, ... As an example of this procedure, axiom 4 of the standard arithmetic has the following Gödel number:

$$x_1 + sx_{11} = s(x_1 + x_{11})$$

$$2^9 \cdot 3^{10} \cdot 5^3 \cdot 7^2 \cdot 11^9 \cdot 13^{10} \cdot 17^{10} \cdot 19^5 \cdot 23^2 \cdot 29^6 \cdot 31^9 \cdot 37^{10} \cdot 41^3 \cdot 43^9 \cdot 47^{10} \cdot 53^{10} \cdot 59^7 \cdot$$

The number was obtained by scanning the given expression one symbol at a time and converting it to the appropriate prime power. Thus x, the first symbol, has code number 9; so 2, the first prime, is raised to the ninth power. The next symbol, 1, becomes 3^{10} because 3 is the next prime and 10 is the code number for the symbol 1.

Note that the axiom has been altered to accommodate our use above of a special notation for variables, namely, to use a sort of unary code (consisting of consecutive 1s) to yield a system of subscripts for the symbol x. In other words, x_1 and x_{11} may be considered as perfectly general names like x and y in axiom 4 of the arithmetic.

As can be seen in the example above, Gödel numbers tend to be enormous. Nevertheless, they are computable, and given any integer whatsoever, it is possible to compute the expression it represents (if any) by finding all its prime factors and grouping these as powers in order of increasing primes.

We are now ready to enter the very heart of Gödel's theorem by considering the following predicate:

$$\text{Proof}(x,y,z)$$

Here, $\text{Proof}(x,y,z)$ is a predicate having the following interpretation: "x is the Gödel number of a proof X of a formula Y (with one free variable and Gödel number y) which has the integer z substituted into it." The "proof X" referred to here may itself be considered a formula for the purpose of having a Gödel number assigned to it.

Note that the basic symbols to which code numbers were attached did not include the predicate symbol "Proof" or any other predicate symbol except equality ($=$). Indeed, "$\text{Proof}(x,y,z)$" is just *our own shorthand* for an immensely long expression with three free variables x, y, and z—or, in Gödel's notation, x_1, x_{11}, and x_{111}. This expression includes a number of procedures, including the following:

1. Given an integer, produce the string of which it is the Gödel number.
2. Given a string, check whether it is a formula.
3. Given a sequence of formulas, check whether it is a proof for the last formula in the sequence.

All these procedures are computable and, as Gödel showed, themselves reducible to formulas within the formal system defined above. Before we show how this predicate is used in Gödel's theorem, there is one small detail to clean up. Namely, we must state what a "formula" really is: The definition embedded in procedure 2 above would amount to an inductive definition of a properly formed arithmetic expression and the ways in which such expressions may be combined legally with the logical connectives & and → or quantified over by ∃ and ∀. For example, $\exists x_1(X_{11}(x) = X_{111}(x))$ is a formula but $1X)\exists x_{11}\daleth =)$) is not.

We are now ready to consider a very special use of the predicate under consideration. Suppose that the formula Y is fed its own Gödel number and that we deny the existence of a proof within the formal system of the resulting formula:

$$\daleth x \text{Proof}(x,y,y)$$

In words, the formula Proof(x,y,y) says, "x is the Gödel number of a proof of the formula obtained by substituting its own Gödel number y for its one free variable." Consequently, to write $\daleth x$ in front of this is to deny the existence of such a proof.

Now any such predicate expressible in our formal system has a Gödel number, and it is amusing to consider the cartoons in Figure 5.1.

In the first instance, we have the single-free-variable expression $\daleth x \text{Proof}(x,y,y)$ preparing to eat y. Its own Gödel number is denoted by g. In the next instance, the character is given its own Gödel number to eat, and upon its ingestion the character is transformed to a predicate with no free variables — and hence no mouth. Naturally, even the resulting formula has a Gödel number, g'.

Gödel's Theorem: $\daleth x \text{Proof}(x,g,g)$ is true but not provable in the formal arithmetic system.

The proof of this theorem takes only a few lines:

Suppose $\daleth x \text{Proof}(x,g,g)$ is provable in the system, and let p be the Gödel number of its proof P. We then have

Figure 5.1 A certain formula ingests its own Gödel number

$$\text{Proof}(p,g,g)$$

is true since P is a proof of G with g substituted for its one free variable. But, of course, $\text{Proof}(p,g,g)$ contradicts $\exists x \text{Proof}(x,g,g)$, and we are left with the conclusion that no such proof P exists.

The formula $\exists x \text{Proof}(x,g,g)$ is certainly true because we have just established the claim which it makes about itself—that it has no proof!

Such a proof, where a two-variable predicate is given the same value for both its arguments, is called a *proof by diagonalization,* and it crops up frequently in the theory of infinite sets and mathematical logic. Cantor was the first to use such an argument in showing that the real numbers are not countable.

Are there true statements which mathematicians are trying to prove at this very moment, but never will? What about Goldbach's conjecture which states that every even number is the sum of two primes? Certainly, no one has proved this statement so far, yet most mathematicians think it is true.

The struggle to formalize mathematics in a mechanical way led to the discovery of a basic and deeply seated problem in mathematics itself. The discovery was to be paralleled a few years later when the attempt to formalize "effective procedures" led to the discovery of a basic inadequacy in computers. There are some tasks just as impossible for computers (see Chapter 59) as for mathematicians.

Problems

1. Write out the Gödel numbers for the integers 0, 1, 2, and 3.

2. Is it possible for two different expressions to have the same Gödel number? If so, give an example. If not, explain the impossibility.

3. What is the difference between our proof of Gödel's theorem and a proof in the formal arithmetic system? Can our proof ever be expressed as a proof in this system?

References

S. C. Kleene. *Introduction to Metamathematics.* Van Nostrand, Princeton, N.J., 1950.

Raymond M. Smullyan. *Gödel's Incompleteness Theorem.* Oxford University Press, New York, 1992.

GAME TREES

The Minimax Method

A standard approach to programming a computer to play a given two-person game is to use a "game tree" in which each node represents a possible position in the game and each branch represents a possible move. The rather simple game of 4×4 checkers provides a suitable illustration of this idea (Figure 6.1).

At each level of the tree, moves alternate between red and black. Suppose now that we had a magical device which was able to analyze the boards at the third level of the tree and arrive at a number which reflected the value to black of each board on that level. For example, on the extreme left-hand board, black is about to lose, the board next to this is a draw, and the board after that is a win for black. These possibilities are reflected in the numbers -1, 0, and $+1$. On the move leading to any of these three boards, it is red's turn, however, and red would obviously select the extreme left-hand move. It follows that the value to black of the left-hand board on the second row is -1 since, having selected that move, black may as well assume that red will make the best possible move against black.

Thus for each board on the second row, one might select the minimum value

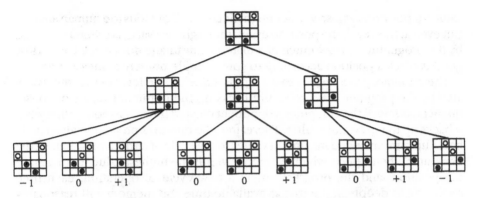

Figure 6.1 Part of the game tree for 4 × 4 checkers

assigned to its consequent boards as the value to black of that board. This leads to the sequence − 1, 0, − 1 of values to black of the three boards on the second level. However, black can choose whichever move it likes to reach this level and so will choose the move yielding the maximum value, in this case the middle move, resulting in a board of value 0 — at least black will not lose!

The process just illustrated is called the *minimax procedure*. A computer program to play black in 4 × 4 checkers could exploit the game tree structure through the following three subprograms:

TREE GENERATOR
POSITION EVALUATOR
MINIMAX PROCEDURE

Tree generators are not hard to construct for most two-person games. Having decided on a method of representing the board (or game situation), the programmer merely designs a procedure for generating and storing all the legal moves from that position.

The position evaluator is a bit trickier. If it were possible to generate all possible outcomes of a game, the position evaluator would have the relatively simple task of

Recognizing repeated positions (a draw)
Recognizing wins
Recognizing losses

Normally, there is neither time nor space to generate the entire game tree, and the position evaluator is called in at a much higher level, being asked to return a

37

value for positions whose winner might not be at all obvious to a human analyst. But even at this level, the position evaluator might use various criteria specified by the programmer: How much of a material advantage does black have? How good is black's position according to some simple numerical measure?

The minimax procedure is easiest to describe. At the level of the tree evaluated by the program, this procedure backs up the minimum values of consequent positions if the computer's opponent created those positions; otherwise, it backs up the maximum value. Thereafter, the minimax procedure alternately backs up minimum and maximum values to the node of current play. Black (the computer) then selects whichever move yields the highest value in the resulting position, and the program enters a new cycle of operation, each time exploring as deeply in the tree as available time and memory will reasonably allow.

Just such a program was written by Arthur Samuel in 1962. This program played normal checkers and once beat a state champion in the United States.

In more complicated games such as chess, with many more moves per turn, the role of the position evaluator becomes even more critical, and only chess-playing programs (most of which use some variant of the basic strategy outlined here) have reached the same level as the Simon program, relatively speaking.

For this reason, it becomes especially valuable, in more complex games such as chess and go, to have some technique for pruning the game tree so that it does not become unwieldy too quickly. Such a technique exists.

Examining the top portion of the 4×4 checkers game tree once again (Figure 6.2), we notice that for two of the positions available to black on the first move, there was a move available to red which resulted in a position value to black of -1. If the only values available to us at the third level were the two -1s,

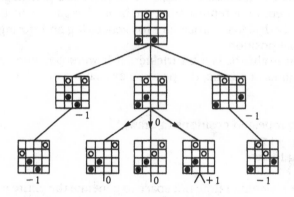

Figure 6.2 Pruning the game tree

+1

Figure 6.3 A consequent board is a win for Black

we could nevertheless eliminate the necessity of exploring the tree further along any of the branches leading from the positions above the −1s. Two branches of our game tree would then be pruned on the assumption that the opponent will select a line of play at the second level that leads to a −1 score.

At this stage there would be only three branches of the game tree to explore, namely, the three middle ones at the third level. Suppose that in exploring the third branch the game-playing program arrived at a value of +1 for one of the two possible consequent boards (Figure 6.3). There would then be no need to explore the other branch since the computer would already have a perfectly good move available to it from the previous board. We assume that red would therefore avoid that board by choosing another move.

This sort of pruning of the game tree is called the *alpha-beta procedure*. Like the minimax procedure, it searches a position value α which represents the smallest value to which black can be held by any move that red might make (Figure 6.4). Specifically, suppose that a red position C is already known to have value α. It may happen that exploration of a branch (E) from a lateral position B results in a value of $v < \alpha$. Then there is no point in exploring the other branches of B since it is already clear that black should prefer move C to move B: from the latter, red can reduce black's score to v. Thus, the branches at B are pruned, and analysis of the game tree moves on to position D.

In searching for a maximum, the alpha-beta procedure looks for the largest value β to which red can be held (Figure 6.5). If it is known that a black position

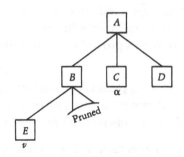

Figure 6.4 An α cutoff

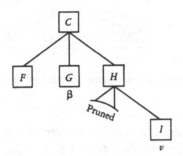

Figure 6.5 A β cutoff

G already has value β and if exploration from H yields a position I with value $v > \beta$, then clearly red will prefer move G to move H and there is no point exploring the other branches at H any further. A "beta cutoff" at C occurs.

By this additional technique, large chunks of the game tree can be removed from further consideration. Among other things, this gives the programmer the option of increasing the speed of the game-playing program or of improving its performance within the same time frame, by allowing it to explore more of the game tree than was previously possible.

It is not difficult to show that the α-β procedure of tree pruning results in exactly the same moves by black as if the pruned branches were actually explored. Most minimax game-playing programs now use the alpha-beta procedure since it almost always results in enormous savings in time and space. Of course, there are hypothetical trees in which the alpha-beta procedure would save no time at all, but these do not seem to occur in practice.

Problems

1. Obtain a large sheet of paper and attempt to complete the game tree for 4 × 4 checkers. Label each bottom board with a +1 (black win), 0 (draw), or +1 (white win), and use the minimax procedure to back these values up to the top node. Can black win?

2. In describing the alpha-beta pruning procedure, it was assumed that red has access to the same board values as black and behaves accordingly. Must the procedure be changed if red selects random moves? If so, how?

3. Most game-playing programs have a "look-ahead factor" which limits the depth to which the program explores the game tree from any given position.

This means that boards at the limit of this depth must be evaluated according to rough measures of Black's probable success with them. Devise such a measure for 4 × 4 checkers which takes into account only the number of pieces, whether they occupy a center square or edge square, and the kind of moves (advance or jump) available to them. How well would an alpha-beta pruning program do with such a board evaluator and look-ahead factor of 2?

References

A. L. Samuel. Some studies in machine learning using the game of checkers. *Computers and Thought* (E. A. Feigenbaum and J. Feldman, eds.). McGraw-Hill, New York, 1963, pp. 71–108.

Patrick Henry Winston. *Artificial Intelligence,* 3d ed., Addison-Wesley, Reading, Mass., 1992.

THE CHOMSKY HIERARCHY

Four Computers

At first glance, a theoretical computing model could not be more different from computing hardware. The former is abstract while the latter is concrete. Yet by stressing only the salient points of computation, a theoretical model enables us to delineate precisely the powers and limitations of a specific computational scheme. Although hundreds of computing models have been proposed since the inception of computer science, each has tended to be equivalent to just one of four principal models: finite automata, pushdown automata, linear bounded automata, and Turing machines. These abstract computers form the subject of four separate chapters — Chapters 2, 14, 26, and 31. In this chapter, the four schemes are viewed as variations on a single theme symbolized by the schematic machine shown in Figure 7.1.

The box embodies a set of rules that determine how the machine will respond to a tape that is fed to it, one cell at a time. Each cell contains a symbol drawn from a finite alphabet. The machine operates in cycles. Each cycle consists of reading a symbol, responding, and then moving the tape. The response of the machine determines, in general, which of the four types it is.

The Chomsky hierarchy is named after the U.S. linguist and philosopher of

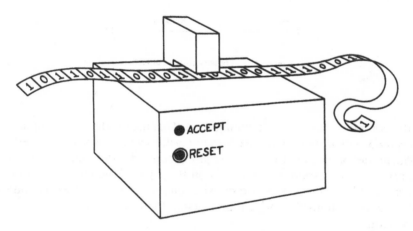

Figure 7.1 The basic machine

language Noam Chomsky. Each model of computation determines a class of languages. Since each model is more general than its predecessor in the hierarchy, each class of languages includes the one before it:

Increasing generality
→

Computing model	Finite automata	Pushdown automata	Linear bounded automata	Turing machines
Language class	Regular languages	Context-free languages	Context-sensitive languages	Recursively enumerable languages

A specific example of one of the four models may as well be called a *computer*. It has a finite number of states, and each cycle of operation involves a transition from one state to another that is triggered by a specific symbol from the computer's alphabet. The states of the computer can be symbolized by labeled circles (Figure 7.2), and the transitions can be represented by arrows between them.

The computer has a special *initial state* and one or more *final* or *accepting states*. When a word x is placed on the tape at the beginning of a computer's operation, the computer is in the initial state. It proceeds to respond to the tape, symbol by symbol. If the "accept" light goes on after the last symbol of x is read

43

Figure 7.2 Transition in an automaton

by the computer, it is in an accepting state. This means that the computer has *accepted x*. The *language* of the computer is the set of all words accepted by it.

The model of the generic computer just described is determined entirely by the nature of its response to symbols on the tape. If the tape moves in one direction only and the response consists merely in the advance of the tape followed by a transition to another state, then the computer is a finite automaton.

There is nothing to prevent the computer, whatever model it might happen to be, from writing on the tape. The presence or absence of such output symbols makes no difference, when the tape is unidirectional, to the language that the computer accepts. Nevertheless, we distinguish this variant of a finite automaton as a *Mealy machine*, after the mathematician G. H. Mealy. In such a model, transitions in the computer would be represented as shown in Figure 7.3. If the computer is in state *i* and the symbol read is *a*, then the computer writes a *b*, advances the tape, and enters state *j*.

The next higher level of generality in the Chomsky hierarchy involves pushdown automata (Figure 7.4). Here, we alter the schematic machine temporarily to include an additional tape.

A computer of this kind is restricted to a unidirectional tape just as a finite automaton is. The auxiliary tape, however, can move in either direction, forward or backward.

Initially, such a computer scans the first cell of the auxiliary tape as well as the first symbol of a word on its main tape. At any point in its computation, a cycle consists of reading the symbols on *both* tapes and responding accordingly: Before advancing the main tape and entering the next state, the computer may

Either advance the auxiliary tape and write a symbol on the next cell
Or erase the symbol in the auxiliary cell currently scanned and move back one cell

Figure 7.3 Transition in a Mealy machine

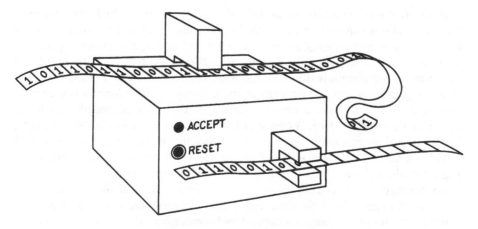

Figure 7.4 A push-down automaton

The first cell of the auxiliary tape contains a special symbol which the computer recognizes as a marker. It is not allowed to write over this marker or to move the tape backward from there; to do so would obviously require it to read a nonexistent cell on the auxiliary tape at the next computational cycle.

In our diagrammatic shorthand (Figure 7.5), a pushdown automaton may thus have two kinds of transition from a given state. The terms *push* and *pop* refer to a more common visualization of the auxiliary tape, namely, as a *pushdown stack*. The symbols on this tape are stacked up in the following sense: To read given symbols on this tape, the computer must read and erase all symbols between *s* and the current scanned cell on the tape. This is akin to removing all the objects on a stack to get at a specific item in the middle.

A symbol *c* that is added to the auxiliary tape is accordingly "pushed" onto a stack. If a symbol is removed from the stack or "popped" therefrom, we do not have to specify which one — it is whatever symbol is currently being scanned.

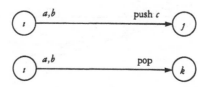

Figure 7.5 Two kinds of transition in a push-down automaton

Pushdown automata have an additional feature not shared by finite automata. Given a current state and pair of symbols on the main and auxiliary tapes, a pushdown automaton is not confined to a specific next state. It may go to one of several (Figure 7.6). Such transitions are called *nondeterministic* since the next state is not uniquely determined by current conditions (see Chapter 26). When it comes to accepting words in its language, a pushdown automaton is assumed always to make a correct transition, that is, a transition that leads toward an accepting state. When it is processing a word not in its language, however, it does not matter how such a computer pushes or pops its stack, or which transition it chooses when offered a choice. It can never be in an accepting state when the last symbol on its input tape is processed.

Accordingly, two kinds of pushdown automata are defined, deterministic and nondeterministic. By definition, the latter class of automata includes the former. The set of languages accepted by (nondeterministic) pushdown automata is called *context-free* (see Chapter 26). Languages accepted by deterministic pushdown automata are called *deterministic context-free.*

Obviously, the class of pushdown automata includes the class of finite automata: Any finite automaton can be made into a pushdown automaton by equipping it with an auxiliary tape which it simply ignores, in effect. It may simply keep writing the same symbol on its auxiliary tape as it reads and responds to symbols on the main tape.

A *linear bounded automaton,* like a pushdown automaton, may operate nondeterministically, but it is not equipped with an auxiliary type as the pushdown automaton is. Instead, it has a single, main tape as the finite automaton does. However, it is allowed both to read from and write upon its tape. As we have already seen in the case of Mealy machines, this does not give the linear bounded automaton any special advantage unless it can move its tape in either direction. The response of a linear bounded automaton can therefore be indicated in Figure 7.7. Here, the computer is in state i and currently scans the symbol a on its tape. It writes the symbol b in place of a and then moves the tape

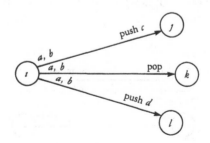

Figure 7.6 A non-deterministic transition

Figure 7.7   A more general kind of transition

in the direction D. In an actual transition, D might be L for left, R for right, or S for stop.

In fact, the computer just defined is a *Turing machine,* the most general computational model known (see Chapters 31 and 39). To complete the definition of a linear bounded automaton, one condition must be added: For any given input word x of length n (containing n symbols), a linear bounded automaton is only allowed a constant k times as much total tape to complete its computation. For a given linear bounded automaton, for example, the constant k might be 5. To decide whether a word of length 7 is accepted, the automaton would thus be allowed the use of 35 consecutive tape cells and no more. Seven of these cells would be occupied by the word, and the remaining 28 cells may be used as a kind of workspace.

Linear bounded automata accept the class of languages called *context-sensitive.* Turing machines accept the class of languages called *recursively enumerable.*

The class of linear bounded automata includes the class of pushdown automata. The proof of this fact is not quite as simple as the proof that the pushdown automata include the finite automata. Given an arbitrary pushdown automaton M, we may, however, easily construct a linear bounded automaton that simulates it. The key to the simulation lies in the fact that a pushdown automaton uses an amount of auxiliary tape (the pushdown stack) that is bounded linearly by the size of the input word: As it processes this word, symbol by symbol, it uses at most a fixed number k of auxiliary tape cells. The linear bounded automaton M' that simulates M uses its workspace to mimic the auxiliary tape. It shuttles back and forth between the input word and workspace, alternately enacting operations on the main and auxiliary tapes.

In the previous definition of a Turing machine, it was implicitly assumed that a Turing machine was nondeterministic. As normally defined (see Chapter 31), Turing machines are deterministic. For each current state and input symbol combination, there is only one response open to this model of computer. Indeed, it makes no difference. One may, with some patience, construct an equivalent deterministic Turing machine for an arbitrary nondeterministic one.

Unfortunately, no one knows whether deterministic linear bounded automata are equivalent to nondeterministic ones.

This chapter has stressed the performance of various styles of abstract computers as language acceptors. In fact, computers are more commonly regarded as input/output transducers. The word on the tape at the beginning of the computation is transformed, when (and if) the computer halts, into a new word. By such words one may represent numbers, computer programs, or virtually any well-defined symbolic entity.

Problems

1. Show that no finite automaton can accept the language consisting of all words of the form $a^n b^n$, $n = 1, 2, 3, \ldots$. The formula represents n a's followed by n b's.

2. Construct a deterministic pushdown automaton that accepts the language of Problem 1.

3. Show that no pushdown automaton, deterministic or nondeterministic, will accept the language consisting of all words $a^n b^n c^n$, $n = 1, 2, 3, \ldots$. Having done this, construct a linear bounded automaton that will.

References

Derek Wood. *Theory of Computation.* Harper and Row, New York, 1987.

Daniel I. A. Cohen. *Introduction to Computer Theory.* Wiley, New York, 1986.

RANDOM NUMBERS

The Chaitin-Kolmogoroff Theory

To mention random numbers and computers in the same breath seems almost a contradiction—the essence of randomness is the absence of procedure or mechanism. Consider the case of the random number machine used in some lotteries (Figure 8.1). Ten colored balls with numbers 0 through 9 circulate in a metal cage. A continuous blast of air keeps the balls aloft. When the air is turned off, one of the balls enters a channel and exits the cage to the outer world. This is a random number. Or is it?

To the extent that the laws of physics uniquely determine the balls' position at each moment, this procedure for getting random numbers is hardly perfect. Lottery winners, however, have never been known to raise philosophical objections to the device.

And rarely do users of most modern computer languages object to the random number generators used by such languages. The "random" numbers thus produced are so useful in various applications (see Chapter 4) that few people give the matter a second thought, provided the numbers *seem* random.

Many of the generators employed in modern computers use the linear congruential method. Here a simple linear formula operates on the present random

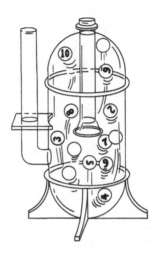

Figure 8.1 A randomizing device for lotteries

number to get the next one:

$$x_{n+1} \leftarrow (k \cdot x_n + c) \bmod m$$

The current random number x_n is multiplied by multiplier k, and an offset c is added to the product. Finally, the resulting number is taken modulo m, the modulus. To start the whole process, an initial number x_0, called the *seed*, is chosen.

Not every combination of parameters k, c, and m is equally effective at producing seemingly random numbers. For example, the following values (drawn somewhat randomly from the author's imagination) produce a sequence that has a major shortcoming:

$$k = 19 \qquad c = 51 \qquad m = 100 \qquad x_0 = 25$$

25, 26, 45, 6, 47, 44, 87, 4, 27, 64, 67, 24, 7, 84, 47, . . .

The last number listed here, 47, is the same as the fifth number in the sequence. Since the formula used is entirely deterministic, the subsequence between the two 47s will go on repeating endlessly. The length of such a subsequence is called the *period* of the sequence. Obviously, every sequence of numbers generated by this method will repeat itself sooner or later, but the question arises of how much better one can do.

The next choice of parameters improves things somewhat. It also illustrates how sensitive the generation process is to small parameter changes:

$$k = 19 \qquad c = 51 \qquad m = 101 \qquad x_0 = 25$$

25, 21, 46, 16, 52, 29, 97, 76, 81, 75, 62, 17,
71, 87, 88, 6, 64, 55, 86, 69, 49, 73, 24, 2, 89, 76, . . .

Here the sequence does not begin to repeat itself until much later. The period has improved from 10 to 18.

Seemingly more random sequences of numbers may be produced by means of the well-known logistic formula used as an illustration of chaos in dynamic systems:

$$x_{n+1} \leftarrow r \cdot x_n \cdot (1 - x_n)$$

By starting with an initial seed value, x_0 lying between 0 and 1, the iterative formula can be made to produce a most convincing-looking sequence of random numbers. The seed value is not critical. After a few iterations, the sequence of values will hop madly about within a subinterval $[a, b]$ of $[0, 1]$. By applying the transformation

$$y_n \leftarrow \frac{x_n - a}{b - a}$$

to x_n, one obtains a new "random" number y_n between 0 and 1.

The logistic formula only behaves chaotically for certain values of the parameter r. Readers might enjoy programming the formula and exploring its behavior from $r = 3.57$ up to $r = 4$.

Much theory has been elaborated to improve the randomness of various generating programs, but they will always be what can only be called pseudorandom. They do not answer the question, What is random? But they do illustrate the idea that computer programs can generate numbers of varying degrees of apparent randomness.

The idea of a computer program generating a sequence of random numbers underlies the Chaitin-Kolmogoroff theory. Discovered independently in the mid 1960s by Gregory J. Chaitin of IBM's Thomas J. Watson Research Center in Yorktown, New York, and by A. N. Kolmogoroff, a Russian mathematician, the theory defines the randomness of a finite sequence of numbers in terms of the length of the shortest program that will generate it; the longer the program has to be, the more random the sequence. Obviously, a program that produces a given sequence hardly needs to be longer than the sequence itself. This suggests that those sequences requiring programs approximately as long as themselves are the most random; it is reasonable to tag these, at least, with the adjective *random*.

Consider the binary equivalent of the sequences generated at the beginning of this chapter. What is the shortest program that generates the sequence consisting of *m* repetitions of 01001? Assuming for the moment that the algorithmic language used in this book is a programming language, one can at least bound the minimum length of a generating program by one that generates this sequence:

for $i \leftarrow 1$ **to** m
    output 0,1,0,0,1

This program contains 23 ASCII characters (not counting blanks) and a variable *m* that will change from one version of the program to the next. In fact, for a particular value of *m*, the program length in characters will be $23 + \log m$. It generates a sequence of length $n = 5m$. How random, then, is the sequence it produces? One way to measure randomness is to form the ratio of program length to string length. The sequence consisting of *m* repetitions of 0, 1, 0, 0, 1 thus has randomness

$$r \leq \frac{23 + \log m}{5m}$$

Since this ratio tends to 0 as *m* gets large, it is reasonable to conclude that longer and longer versions of the sequence are less and less random. In the limit, this ratio has randomness 0; in other words, it is not random at all. The same conclusion will hold for any program that has a fixed number of parameters related to sequence length; the numbers it produces will tend to have no randomness.

Over all sequences of length *n*, however, it can be shown that the vast majority are random. By using the arbitrary threshold of $n - 10$, it may be asked how many *n*-digit sequences are generated by minimal programs of length less than $n - 10$. For the sake of argument, we assume that all programs are written in terms of binary digits. This does no harm since alphabetic and other symbols can be regarded as 8-bit groups of digits. Altogether there are certainly no more than

$$2^1 + 2^2 + \cdots + 2^{n-11}$$

programs of length less than $n - 10$. The sum does not exceed 2^{n-10}. It follows that fewer than 2^{n-10} programs have length less than $n - 10$. These programs generate at most 2^{n-10} sequences, and the latter account for about one n-bit sequence in every thousand!

One would think that with so many random sequences about, it must be an easy matter to display one. Nothing could be further from the truth. To prove that a particular sequence S is random, one must show that no program significantly shorter than S will generate it.

Suppose that such a proof process itself has been mechanized in the form of a program P. The program operates on statements in the predicate calculus (see Chapter 58) and, for each one, decides whether it is proof that a particular sequence of n bits can only be generated by a program as long as the sequence. Actually, P does not need to be quite as general as this; it only needs to check that the sequence can only be generated by a program longer than P.

There is a mechanical procedure for generating valid predicate formulas, one after another, in such a way that their length steadily increases — all the formulas of length 1, then all the formulas of length 2, and so on. Some of the formulas will turn out to be proofs that specific sequences cannot be generated by programs that are as short as P. But in such a case, P could be made to report such sequences, generating them, in effect. Thus P will have generated a sequence that it is too short to generate.

The apparent contradiction forces us to conclude that P cannot exist. Neither, then, can any program (or proof procedure) more general than P. Thus it is impossible to prove that a given sequence is random in spite of the fact that most sequences are random!

The resemblance between the argument just given and Gödel's famous incompleteness theorem (see Chapter 5) is not accidental. In the type of formal system defined by Gödel, it has been shown that any axiom system for it is incomplete — there are theorems that cannot be proved in the system. Among such theorems are those asserting that a certain long sequence of numbers is random. The bottom line of the Chaitin-Kolmogoroff theory touched upon here is that while we shall never know for sure whether a given sequence is random, we can at least measure the degree of randomness in sequences produced by programs.

For those who might despair of ever producing a sequence of truly random numbers, there is a device that does the job. A zener diode is an electronic component that passes current in one direction. When it is operated under reverse voltage, some electrons leak through the device in the wrong direction. Their frequency depends on random thermal motions of electrons inside the diode. When it is measured on a sensitive oscilloscope, the leaking current certainly looks random (Figure 8.2).

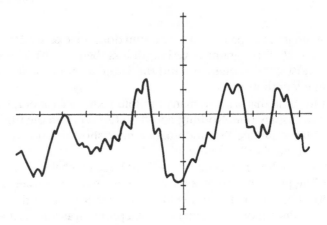

Figure 8.2 Electronic noise

It is possible to incorporate the diode into a circuit that samples the current at regular intervals a few microseconds apart. If the value exceeds a certain threshold, an analog-to-digital converter generates a 1 bit; if not, it generates a 0 bit. In this way a sequence of numbers is produced. If the sequence is not random, then modern quantum mechanics is in trouble.

Problems

1. Find a combination of k and c that, with $m = 100$, produces a random-looking sequence with a period of 50.

2. Write the shortest algorithm you can think of to generate the sequence 0, 1, 0, 0, 1, 0, 0, 0, 1, . . . up to some length n. Use the algorithm to derive an upper bound for the randomness of the sequence.

3. Show that a sequence that is random by the Chaitin-Kolmogoroff definition must have an approximately equal number of 0s and 1s in the long run.
   (*Hint:* Assume that the sequence contains n symbols, z of which are zeros. If the ratio $r = z/n$ is relatively small, display an algorithm that is in the order of r times the length of the sequence.)

References

D. E. Knuth. *The Art of Programming,* vol. 2. Addison-Wesley, Reading, Mass., 1967.

Gregory J. Chaitin. *Algorithmic Information Theory.* Cambridge University Press, Cambridge, 1987.

MATHEMATICAL RESEARCH

The Mandelbrot Set

The advent of the computer has spurred the development of many new fields of mathematics, not to mention considerably advancing some older fields. One has only to think of the computer proof of the four-color theorem.

Many mathematical objects, from systems of differential equations to complicated topological groups, can be explored not only in the mind but by computer. Sometimes it is possible to visualize the object on a computer screen. This is the case with the famed Mandelbrot set (Figure 9.1). Rarely has a mathematical object created so much excitement in the wider world.

The story of how mathematicians regard the Mandelbrot set has yet to be told publicly. In fact the Mandelbrot set was first defined in 1905 by Pierre Fatou, the great French mathematician. He developed the field now known as complex analytic dynamics. Fatou studied iterative processes with formulas like

$$z = z^2 + c$$

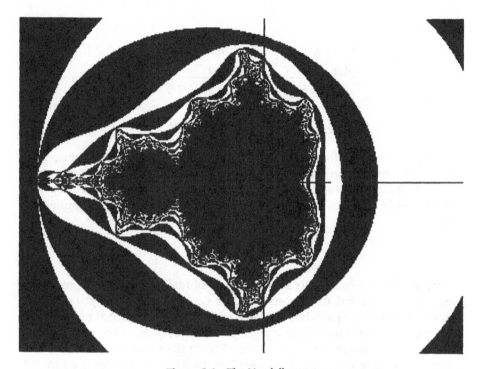

Figure 9.1   The Mandelbrot set

in which a complex number, z, is given some initial value and then is iterated through the equation. The complex number c, on the other hand, remains constant.

The formula gives rise to a succession of z-values in the following manner. Starting with the initial value, z_0, each value is substituted into the expression on the right-hand side of the equation and the expression is then evaluated. The resulting value for the expression is then substituted for z in the right-hand side of the equation used and the process is repeated ad infinitum. Mathematicians call the sequence that results the *orbit* of z_0.

Fatou was especially interested in the orbit of $z_0 = 0$ for different values of the constant c. Was there some regularity that he might detect and exploit?

Complex numbers are actually pairs of real numbers written in algebraic form. The complex number z, for example, may be written as a formal sum of two real numbers $x + iy$, where i is the so-called *imaginary* number that represents the square root of -1. Complex numbers may be added and multiplied. For example, here are the complex numbers $3 + 7i$ and $8 - 5i$ added and multiplied:

$$
\begin{array}{lr}
& 8 - 5i \\
\text{plus } \underline{3 + 7i} \\
11 + 2i
\end{array}
\qquad
\begin{array}{lr}
& 8 - 5i \\
\text{times } \underline{3 + 7i} \\
56i - 35i^2 \\
\underline{24 - 15i} \\
24 + 41i + 35 = 59 + 41i
\end{array}
$$

The addition is straightforward enough, the numbers being added component-wise. But the multiplication resembles an exercise in high school algebra. It all hinges on the fact that the square of i is -1. The complex version of 0 is $0 + 0i$ and the size of a complex number can be expressed in a formula that applies to z above.

$$|z| = \sqrt{x^2 + y^2}$$

Since a complex number is actually a pair of real numbers, it represents a point on a plane called, not surprisingly, the *complex plane*. Not just complex numbers with integer parts but complex numbers with fractional or even irrational parts populate this plane and fill it completely. In other words, every point on the plane corresponds to a complex number and vice versa.

It therefore becomes possible to visualize the iterative process in a formula such as the one Fatou played with by plotting the orbit of complex numbers there. Depending on the value of c, the orbit would sometimes bounce back and forth among a few points, sometimes move about the origin in an erratic manner, and sometimes grow without limit, apparently heading for infinity. In the first two cases the numbers always remained bounded in size, that is, in distance from the origin (Figure 9.2). In the latter case, they got larger and larger without limit. Fatou proved some interesting things about the "dynamics" of these orbits.

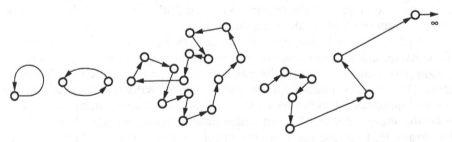

Figure 9.2 Bounded orbits (left three) and unbounded orbit (right)

Still fundamental today is a simple lemma of Fatou's: If the size $|z|$ of the number z ever exceeds 2, the orbit is destined for infinity. This made an interesting and useful test for deciding the destiny of the dynamics for a given complex constant c. Simply iterate the formula, starting at $z = 0$, and if the value or modulus of z ever exceeds 2, stop.

There is a problem with this rule, of course. Even if an orbit is destined for infinity, it might take a very long time to exceed 2. So long, in fact, that a tired Fatou might feel the issue in doubt and cease his laborious hand calculations. The availability of computers hardly improves things as far as this particular problem is concerned, however. For many orbits we can decide the issue much faster than Fatou could, but an infinity of tough cases will always remain.

Fatou defined the set Kc of all complex numbers that have bounded orbits. He proved a famous result that is still key to the field today: First, he proved that either the set Kc is connected in the topological sense or it is a Cantor set—totally disconnected—a "dust," as some mathematicians call it. Then he proved that the set Kc is connected if and only if it contains the origin, the complex number $0 + 0i$. In other words, Kc is connected if and only if the orbit of 0 is bounded! This amazing result led naturally to a definition of a new set consisting of all the values of C for which Kc was connected: the Mandelbrot set.

Fatou, living long before the advent of the computer, had only the haziest notion of what this set might look like when plotted on the plane. He can hardly have suspected the rare beauties and strange landscapes hidden within it. The American mathematician John Hubbard and the French mathematician Adrien Douady studied the set extensively in the 1970s and early 1980s. Hearing that Benoit Mandelbrot had been producing computer images of the set, they wrote and requested some pictures.

The images were fascinating. There was not only the Mandelbrot set, floating in all its majesty in the middle of the complex plane, there were apparently little Mandelbrot sets (mini-Mandelbrots, as Hubbard calls them) floating around the larger body. A question immediately arose: Were the sets all connected in one piece or were the mini-Mandelbrots truly detached from the parent body? This question, spurred by their first glimpse of the set, led to a now famous result: The Mandelbrot set is connected!

Apart from the sometimes abstruse properties of the Mandelbrot set investigated by mathematicians, there are its obvious charms: It is a fractal, that is, an object with a fractional dimension. Fractals have details at all levels of magnification and the Mandelbrot set is no exception. In 1980 only a handful of people had ever seen an image of the Mandelbrot set. By 1985, several thousand people had written Mandelbrot programs for themselves, using such programs to explore the set, to magnify any part of it they wished, then to magnify part of the image so produced. Since it is a fractal, one moreover that never quite repeats itself, there is not only detail but *new* detail at every level of magnification.

Any readers with only modest programming skills may write their own program for generating pictures of the Mandelbrot set. The basic algorithm involves generating a succession of c-values that represent a grid of points inside the particular area of the set to be examined. For each such value of c the basic formula is iterated. At each iteration the algorithm tests whether the size of the variable z exceeds 2 or not. If it does, the iteration is ended immediately. If not, it continues to the full limit of 100 repetitions.

When the iteration for a given c-value is complete, a color is assigned by the algorithm to the corresponding point on the screen: black if the full limit of 100 iterations has been tried, some other color if not.

MANDELBROT

1. **input** *acorn, bcorn, size*
2. **for** $j \leftarrow 1$ **to** 150
   1. **for** $k \leftarrow 1$ **to** 150
      1. *count* $\leftarrow 0$
      2. *ca* $\leftarrow$ *acorn* $+ j^*$size/150
      3. *cb* $\leftarrow$ *bcorn* $+ k^*$size/150
      4. *zx* $\leftarrow 0$
      5. *zy* $\leftarrow 0$
      6. **repeat**
         1. *count* $\leftarrow$ count $+ 1$
         2. *xtemp* $\leftarrow x^2 - y^2$
         3. *zy* $\leftarrow 2^*zx^*zy + cb$
         4. *zx* $\leftarrow$ *xtemp* $+ ca$
      7. **until** *count* $= 100$ or $x^2 + y^2 > 4$
      8. **color** $\leftarrow$ function(count)
      9. **display** *jk*th pixel with color
   2. **end** for k
3. **end** for j

The first step of the algorithm specifies that a user must enter three numbers into the computer. The variables *acorn* and *bcorn* are the x- and y-coordinates of the upper left corner of the area to be examined. The variable *size* specifies the width of the area. For very high magnifications, when a very tiny part of the mandelbrot set is to be examined, the value entered for *size* will be very small.

Assuming that the area to be computed will take up a square part of the screen that is 150 pixels wide by 150 pixels high, the double loop at lines 2 and 2.1 (see Chapter 1 for guidance on algorithmic notation), set up the grid of pixels to be processed. The grid is indexed by the variables *j* and *k*. The first step within the grand double loop is to set a counting variable called *count* to 0. The two parts of the complex number *c* are then computed in the form of *ca* and *cb* and the complex number *z* is set to 0 by initializing its two components, *zx* and *zy*, to 0.

At this point all is in readiness for the inner loop to compute the orbit of 0 for the *c*-value just set up. Will the orbit remain bounded or will it scream off to infinity? First *count* is incremented, then the *x*-part of *z* is computed in the form of a temporary variable, *xtemp*. After all, the value of *zx* must not be changed until after line 2.1.6.3, where *zy* is computed. In this line and the next, the real and imaginary parts of *c* are added to *zy* and *zx* in that order. In the last line of the inner loop, 2.1.6.4, it is safe to move the temporary value of *zx* into zx.

The inner loop repeats the four steps over and over until either *count* has reached 100, in which case the program assumes (not always correctly) that the orbit was never destined for greater things, or the size of *z* exceeds 2. In the latter case, the algorithm adopts the arithmetically more efficient test of whether the sum of the squares of *zx* and *zy* exceeds 4. It comes to the same thing. Naturally, a program based on this algorithm would save these squares after using them in line 2.1.6.2 so that it could reuse them for the inner loop termination test.

At line 2.1.8 the color of the *jk*th pixel is computed as a function of *count*. Many different possibilities present themselves here. Traditionally, Mandelbrot display programs give the color black to any pixel for which *count* = 100. If *count* is less than 100, however, one may (a) alternate two colors as *count* is found in the range 1 to 5, then 6 to 10, and so on, or (b) assign colors from the screen palette in some order for a specified set of ranges.

The latter option embraces enough possibilities to keep thousands of computers producing images of the same area of the Mandelbrot set for a hundred years, some of the images being more beautiful than anything yet seen! Herein lies whatever artistry one can bring to a computer rendering of the set.

The final line of the algorithm (apart from the loop terminators) is a statement that orders the computer to display the *jk*th pixel with the color computed in the previous step. To program this particular step, the computer's screen coordinates must be used. For example, if the screen is 250 pixels wide and 200 pixels high, and if the origin is in the upper left-hand corner, one would want the center of the square to be close to the center of the screen (125, 100). Thus the program would have to carry out a translation from the index values for *j* and *k* into screen coordinates, *j* being translated from the range (1, 150) to the range (50, 200), for example, This can be done very simply by adding 49 (or 50, it doesn't much matter) to *j* in order to obtain the horizontal screen coordinate for the *jk*th pixel. The vertical coordinate would then be obtained by adding 25 to the *k*-index.

The first thing to be noticed about a working program is that it takes a long time, in general, to compute a complete 150-by-150 image. The program can be speeded up by compiling it first, if this hasn't already been done. It can also be speeded up by reducing the iteration limit from 100 to 75 or even 50. Few people can tell the difference between a Mandelbrot set generated with different limits. Unfortunately, when explorers take an extended cruise into the set, they discover that some parts of it, especially just outside the set where the colors all reside, require much higher iteration limits and a reassignment of colors to resolve the detail.

To get even faster images, programmers may have to resort to assembly code (see Chapter 48), special arithmetic processors, or even parallel machines, now increasingly available. At all events, with a working Mandelbrot program, an unending universe of detail, beauty, and great significance opens before the student of computing. It is even possible that some noticed regularity leads to a new conjecture about the set, one that will keep the practitioners of complex analytic dynamics busy for a while.

Problems

1. Verify that the algorithmic steps 2.1.6.2, 2.1.6.3, and 2.1.6.4 realize the formula

$$z \leftarrow z^2 + c$$

2. Write a program based on the algorithm provided in this chapter and investigate the connectedness problem studied by Douady and Hubbard. Do this by searching for connecting pathways between any of the mini-Mandelbrots and the larger body. How much of any pathway can be seen?

References

H.-O. Peitgen, P. H. Richter. *The Beauty of Fractals.* Springer-Verlag, Berlin, 1986.

A. K. Dewdney. *The Magic Machine.* W. H. Freeman, New York, 1990.

10

PROGRAM CORRECTNESS

Ultimate Debugging

The process of debugging a program sometimes seems to go on forever. This is especially true of programs that were not analyzed before they were written or were not well structured when they were written. Just when a program seems to be running correctly, a new set of inputs results in obviously wrong answers. Program correctness is an issue of increasing importance in a world increasingly reliant on the computer. For some application programs it is not enough to be "sure" that they run correctly; one must *know* it.

Proving the correctness of programs is a discipline which evolved from the pioneering work of C. A. R. Hoare at Oxford University. One makes certain assertions about what a program should have accomplished at various stages of its operation. Assertions are proved by inductive reasoning, sometimes supported by additional mathematical analysis. If a program is proved correct, one may have confidence in it—at least if one's proof is correct. If the proof fails, it may be due to a logical bug in the program which the attempted proof helps to pinpoint. In either event, understanding of the program is often greatly enhanced by attempting a proof of its correctness.

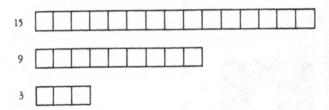

Figure 10.1 Two integers and their greatest common divisor

We illustrate a proof of correctness for the euclidean algorithm (in the form of a program). This algorithm finds the greatest common divisor (gcd) of two positive integers. Figure 10.1 shows two integers, 9 and 15, represented by horizontal bars divided into units.

The gcd of 15 and 9 is 3. In other words, 3 is a common divisor of 15 and 9; it is also the greatest common divisor. The 3 bar, used as a kind of ruler, yields an integral measure for both the 15 bar and the 9 bar. It is also the longest ruler having the property. More than 2000 years ago, Euclid may have used a representation somewhat like this to discover the algorithm named after him. Consider, for example, the two number bars shown below. The 16 bar will fit just once into the 22 bar, and we get a 6 bar left over. Now, compare the 16 bar with the 6 bar. In this case, we can fit two 6 bars into the 16 bar and end up with a 4 bar. We repeat the process. At last a 2 bar is left, and if we now compare a 2 bar with a 4 bar, nothing is left over after the fitting process. Moreover, 2 is the gcd of 22 and 16.

This simple example illustrates the euclidean algorithm. At each stage we take the larger number modulo the smaller until we end up with a zero remainder. In the form of a BASIC program we could write

EUCLID

```
input "M and N, M < N"; M, N
while M > 0
    L = N mod M
    N = M
    M = L
wend 'End of while loop
print N
end
```

A proof that this program produces the gcd of any two positive integers input to it is easiest to illustrate if we represent the program in flowchart form (Figure 10.2).

The flowchart has been tagged with an assertion about the values of M and N input to the program: Both are positive integers, and $M < N$. The program contains a loop, and because the values of L, M, and N change with each

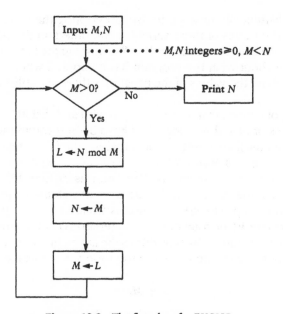

Figure 10.2 The flowchart for EUCLID

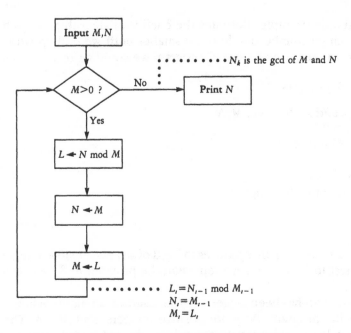

Figure 10.3 The tagged flowchart for EUCLID

iteration, we distinguish these values by subscripting the variables: Denote by L_i, M_i, and N_i the values of these variables computed on the ith iteration, and suppose there are k iterations altogether. We now tag the flowchart with further assertions, one about what the program has computed when it exits from the loop and one about the values of the program's three variables within the loop (Figure 10.3).

We assume, of course, that $M > 0$ to begin with and that M_0 and N_0 represent the input values of M and N before they have been through the first iteration. The tags on the flowchart assert not only that N_k, the last computed value of N, happens to be the gcd of M and N, but also that certain relations hold among the intermediate values of L, M, and N. The latter assertions are obviously true, being mere restatements of what the assignment statements mean in the context of the **while** loop. The first assertion is what we seek to prove.

In the case where k (the total number of iterations) is 1, the program outputs N_1 which happens to equal M_0 according to the tag on the loop. But M_1 must be 0 for the loop not to be reentered. From the loop assertions it follows that

$$L_1 = M_1 = 0$$

whence

$$N_0 \bmod M_0 = 0$$

In other words, N is a multiple of M, and the gcd of M and N must be M itself. But this is simply N_1.

If $k > 1$, then the loop will have at least two successive iterations, and we are in a position to use the following result, put in the form of a lemma. Here $a|b$ means "a divides b evenly."

Lemma: Given two successive iterations of the loop, the ith and $(i+1)$th, a positive integer p satisfies $p|M_i$ and $p|N_i$ if and only if p satisfies $p|M_{i+1}$ and $p|N_{i+1}$.

Proof: According to the loop assertions, $M_{i+1} = L_{i+1} = N_i \bmod M_i$. It follows that there is a positive integer q such that

$$N_i = qM_i + M_{i+1}$$

If $p|(M_i \text{ and } N_i)$, then $p|M_{i+1}$ according to this equation. Also, $N_{i+1} = M_i$, and so $p|N_{i+1}$. However, if $p|(M_{i+1} \text{ and } N_{i+1})$, we may again use the fact that $N_{i+1} = M_i$ to obtain $P|M_i$. But now it is clear from the equation that $p|N_i$, and the lemma is proved.

The chief implication of this lemma is that

$$\gcd(M_i, N_i)|\gcd(M_{i+1}, N_{i+1})$$

and conversely. Hence we have

$$\gcd(M_i, N_i) = \gcd(M_{i+1}, N_{i+1})$$

According to the loop assertions, $N_k = M_{k-1} = N_{k-2} \bmod M_{k-2}$. Moreover, to terminate the loop, it must have been the case that $M_k = 0$ so that $L_k = 0$ and $N_{k-1} \bmod M_{k-1} = 0$. This means that $M_{k-1}|N_{k-1}$. But $L_{k-1} = M_{k-1}$, and L_{k-1} is obviously the gcd of M_{k-1} and N_{k-1}.

Now let L be the gcd of M and N. As the initial step in our simple inductive proof, we note that

$$L = \gcd(M_1, N_1)$$

Suppose that upon completion of the ith iteration, we have

$$L = \gcd(M_i, N_i)$$

Then, by the conclusion drawn from our lemma,

$$L = \gcd(M_{i+1}, N_{i+1}) \qquad i < k$$

The result obviously holds right up to $i = k - 1$, but in this case we have already shown that $L_{k-1} = \gcd(M_{k-1}, N_{k-1})$ and clearly $L = L_{k-1}$. If we recall that the output value of N_k is equal to L_{k-1}, the result follows.

Normally, in proofs of correctness, one also proves that a program terminates for all inputs of interest. In the case of the EUCLID program, one would prove termination by noting, in effect, that M decreases with each iteration and since M is initially a positive number, it must eventually reach zero. Here, the nonnegativity of both M and N plays a role.

For programs with more than one loop, especially when the loops are nested, a somewhat more comprehensive strategy of correctness proving is necessary. In general, besides the assertions of initial and final values, at least one assertion must appear for every closed loop in the program. It is then necessary to prove for each computational path between two adjacent assertions A and A' that if A is true when execution reaches that point, then A' will be true when execution arrives there. This requirement was obviously met in the case of our correctness proof for the single-loop program EUCLID.

Problems

1. Using the same techniques described here, provide a proof that EUCLID terminates for all inputs satisfying the initial conditions.

2. Write a program corresponding to the MINSPAN algorithm described in Chapter 22, and prove that it finds a spanning tree.

3. The assertions used within a program loop are to some degree a matter of taste. Adapt the proof given here to include the additional assertion that

$$\gcd(M_i, N_i) = \gcd(M_{i+1}, N_{i+1})$$

with appropriate restrictions on i.

References

R. B. Anderson. *Proving Programs Correct.* Wiley, New York, 1979.

P. Berlioux and P. Bizard. *Algorithms: The Construction, Proof, and Analysis of Programs* (Annwyl Williams, tr.). Wiley, New York, 1986.

SEARCH TREES

Traversal and Maintenance

One of the most useful data structures ever invented is the tree. Indeed, there are many algorithms for searching and manipulating the data stored in trees. This chapter describes one search algorithm, a few schemes for traversing the tree, and a technique for maintaining (adding to or deleting from) the tree.

Fundamental to the construction of a search tree is the use of nodes and pointers. A *node* is nothing more than a collection of memory locations associated together by a program. Each node has a name which either directly or indirectly refers to the memory address of one of its words. It usually has a content, the item to be stored at the node. It includes, moreover, zero, one, or two (sometimes even more) *pointers,* or *links.* These are nothing more than the names of other nodes in the tree. Programs that use search trees have the option of traveling from one node to another by following such pointers.

Conceptually, a search tree is represented by boxes and arrows (Figure 11.1). Each box represents a node, and each arrow represents a pointer. Each node in the diagram above consists of *fields,* in this case, one data field and two pointer fields.

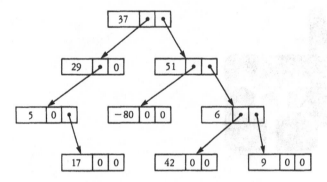

Figure 11.1 A search tree that has nine nodes

A search tree can be implemented directly in main memory by an assembly-level program, or it can be implemented in array form by a high-level program. Figure 11.2 illustrates both possibilities for the tree shown above.

It is not difficult, in either diagram, to trace the data items and pointers from one location to the next; merely start at the arrow in either case. In the first case, the three fields are stored in a single word at memory address 105. The data field contains 37, and the pointer fields indicate the left and right "children" of that node. The left child is at address 102, and the right child is at address 110. In the second example, the data and pointer fields are contained in three separate

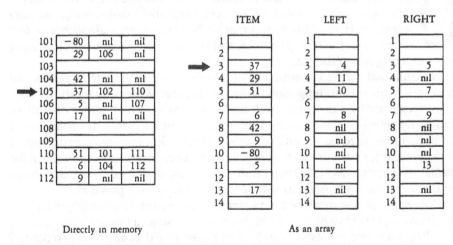

Figure 11.2 Two ways of programming a search tree

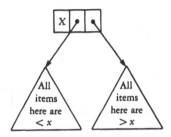

Figure 11.3 A search tree may be organized by an order

arrays called ITEM, LEFT, and RIGHT, respectively. In this case, the fields are used in precisely the same way. Pointer fields containing "nil" do not point to anything, "nil" merely being some symbol which the program can distinguish from a valid pointer address.

Gaps in either representation are shown to stress that trees are sometimes constructed in ways that the programmer cannot always control.

When data are stored in a tree, it is often for the purpose of being able to get to particular items in a hurry. For example, if the items being stored have a special ordering, say < (less than), then the search for items may go especially quickly if they are arranged in the nodes according to the following rule:

At each node, all the items stored in its left subtree are less than the item at that node; all the items stored in the right subtree are greater than the item stored at that node (Figure 11.3).

A data tree satisfying this condition is called a *binary search tree,* and from Figure 11.4, it should be clear how quickly such trees can be searched. In this case, the items are names, and the less-than ordering is alphabetical.

To search this tree, we may use the following simple algorithm. It takes a "name" as input, and it outputs yes or no depending on whether the name was found in the tree. The notations *item (node), left (node),* and *right (node)* refer, respectively, to the item or either of the two pointers at a given node. The algorithm starts at the top node of the tree:

SEARCH

1. *found ← false*
2. *node ← top*
3. repeat

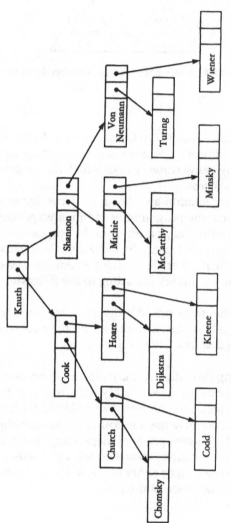

Figure 11.4 Names may replace numbers as data items

1. if *item(node)* = *name*
    then *print* "yes";
           node ← *nil;*
           found ← *true;*
2. if *item(node)* > *name*
    then *node* ← *left (node);*
3. if *item(node)* < *name*
    then *node* ← *right(node);*
4. until *node* = *nil*
5. if not *found*
    then *print* "No".

The procedure SEARCH descends from node to node: If the name sought is less than the name stored at that node, the algorithm takes the left branch; if the name sought is greater than the name at that node, the algorithm takes the right branch. Finally, if the name sought is found at a node, the algorithm prints "Yes."

The loop terminates as soon as *node* = *nil*. This happens not only when the name sought is found, but also when the algorithm tries to descend along a nil pointer — it has reached the bottom of the tree. Note the use of a logic variable *found* which acts as a flag to test, at the end of the algorithm, whether the search was successful. This variable is initially set to *false*.

Clearly, the number of iterations in SEARCH's main loop is never greater than the number of levels in the tree. Since the amount of work done within the loop is bounded by a constant number of steps, we may write the time complexity of search as bounded by $O(l)$, where l is the number of levels in the tree (see Chapter 15. However, in a binary search tree with n nodes and all but the bottommost nodes having two children, the number of levels l is given by

$$l = \lceil \log_2 n \rceil$$

This enables us to write the time complexity of SEARCH as $O(\log n)$ in the general case — at least when the search tree is "full" in the sense given above.

This search time is exceedingly fast. If instead of 15 of the most famous computer scientists, we put 1 billion of the most famous computer scientists,[*] the time would increase from 4 steps to only 30.

From time to time, one might want to list all the data that are stored in a search tree. An algorithm to produce such a list might traverse the tree (visit every node) in one of several orders, the most common being depth-first order. This

---

[*] Among whom the author feels safe in claiming a place.

mode of visitation is easily programmed by using recursion (see Chapters 24 and 55). The algorithm moves up and down within the tree following pointers. The simple rule that defines its progress requires that no nodes to the right, say, of a given node be visited until all the nodes to the left have been visited.

procedure *DEPTH(x)*
        use *x*
        if *left(x)* ≠ *nil*
          then *DEPTH(left (x))*
        if *right(x)* ≠ *nil*
          then *DEPTH(right(x))*

program   *DEPTH(root)*

Here, to "use" *x* means to print it or to process it in some way. If the procedure has not reached the bottom of the tree where pointers are nil, it calls itself first on the left descendant of the current node, then on the right. Of course, the call to DEPTH on the left node might immediately result in another such call and so on until the bottom has been reached. In this way the algorithm, consisting of a single call to DEPTH at the root (or top) of the tree, results in a systematic left-to-right sweep through the tree.

In many applications, binary search trees are not fixed once and for all but grow and shrink as real-world data files do. To insert a new item in a search tree (Figure 11.5), it is only necessary to modify the algorithm above very slightly: As soon as the search for the item fails, get a new (unused) address name *X* and replace the nil pointer just encountered with that address. The next step is to place the item at that address and to create two new nil pointers there. To make a deletion is slightly more complicated. Again, the algorithm above may be used to locate the item, but the address of the node pointing to it must be preserved by the deletion algorithm. In Figure 11.6, a deletion algorithm is shown in operation.

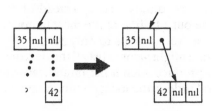

Figure 11.5 Adding a node

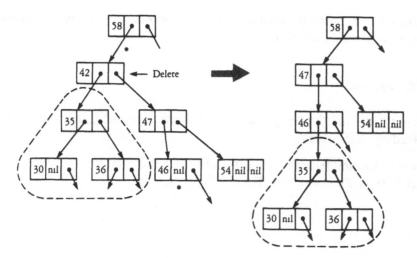

Figure 11.6 Deleting a node

By inspecting Figure 11.6, it becomes clear that the node containing 42 is removed from the tree along with its two pointers. Consequently, two other pointers (marked by an asterisk) have had to be rearranged. Essentially, the whole right subtree pendant at the node containing 42 has been moved up one level while the whole left subtree has slid down two levels to the first available leftmost node of the first subtree. By "available" we mean that the left pointer of that node is nil and one can append the second subtree there.

Problems

1. If a binary search tree has a number of nodes added to it by the method described in this chapter, it does not necessarily stay either full or balanced. In terms of the time it would take to retrieve data from such a tree, what is the worst shape for it to have? Develop an expression for the average time it takes to search an arbitrary binary tree, assuming that all items are equally likely to be searched for.

2. Write insertion and deletion algorithms based on the techniques indicated in Figures 11.5 and 11.6. What is their time complexity?

3. Write an algorithm that processes the nodes of a tree in breadth-first order. This means that nodes are used one level at a time from left to right. Other nodes must, of course, be visited to get from one node at the current level to the next node at the same level.

References

Donald E. Knuth. *The Art of Computer Programming,* vol. 1. Addison-Wesley, Reading, Mass., 1967.

Alfred V. Aho and Jeffrey D. Ullman. *Foundations of Computer Science,* Chapter 5. W. H. Freeman, New York, 1992.

12

ERROR-CORRECTING CODES

Pictures from Space

A space probe speeds toward a distant planet. Within a few days, it will pass near the planet's surface, take several hundred pictures, and record these internally. Even as it zooms off into the depths of space beyond the planet, it begins to transmit its pictures to receiving stations (as in Figure 12.1) on earth across the vastness.

Each picture is divided into horizontal scans, rather like a television image, and each scan line is divided into pixels (individual picture elements) represented by one of 32 possible gray levels — white is symbolized by 0, black by 31, and intermediate values of gray by 1 to 30.

Each pixel is encoded into an electronic message and hurtles at the speed of light back to earth. Even at this speed, the signal can take many minutes to reach the earth, and during its journey it can get garbled. Even when it reaches the receiving antenna, the signal's extreme faintness may cause it to be confused with background noise.

So when an individual pixel, encoded in an electronic pattern of 0s and 1s reaches the receiving station, its original form, say

0101101001011010010110100101011010

Figure 12.1 Telemetry receiving station

might have changed to

$$01001010000101100111101101111010$$

where the arrows indicate errors that have occurred in the message. One might
think that 5 bits of information would suffice to specify one of 32 possible gray
levels. Yet the string displayed above has 32 bits, not 5. It is part of a code
designed to make so many errors correctible.

In fact, the string was part of the code used in the Mariner series of Mars
reconnaissance vehicles. This code is essentially the (32, 5) Reed-Muller code,
defined as follows:

Let
$$H_1 = \begin{bmatrix} 1 & 1 \\ 1 & -1 \end{bmatrix}$$

and define
$$H_{n+1} = H_n \otimes H_1$$

where $\otimes$ denotes the cartesian product of matrices: replace each $+1$ entry of H_n
by H_1 and each -1 entry by $-H_1$. These equations define *Hadamard matrices*.

$M_5 \;=\;$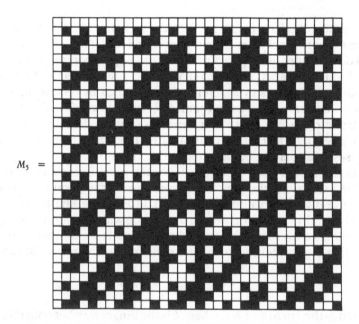

Figure 12.2   The Reed-Muller code as a black-and-white pattern

The Mariner telemetry code consists of the rows of M_5, obtained by replacing 1 by 0 and -1 by 1 in H_5. Matrix M_5 is effectively displayed as a square grid in which 1s are represented by black squares and 0s by white ones, as shown in Figure 12.2.

The rows of M_5 represent the 32 possible code words to be transmitted by the Mariner spacecraft, and they have a very interesting property: Any two differ in exactly 16 places. For example, if we compare the second and third rows, we find that in the following positions one row has a 0 while the other has a 1, or vice versa.

$$2, 3, 6, 7, 10, 11, 14, 15, 18, 19, 22, 23, 26, 27, 30, 31$$

When a new 32-bit code word is received, it is compared by computer to the rows of M_5 and the row it most closely resembles is selected as the word transmitted. Naturally, if there has been only one error in a word, it will differ from one of the rows of M_5 in only one digit and cannot be mistaken for any other row. If two errors have occurred, the same thing is still true. In fact, up to seven errors may occur with no danger of confusion about which word was transmitted by the spacecraft.

But if eight errors have occurred, then the received word may differ as much from some other row of M_5 as from the intended row. This is due to the fact that those two rows differ in only 16 places, and the eight errors might have resulted in a word which is just as "close" to an incorrect row of M_5 as to the correct one.

The number of places in which two binary words or vectors differ is called the *Hamming distance* between them. A set of codewords which are all mutually at Hamming distance d or more enables users to detect and correct up to $\lfloor(d-1)/2\rfloor$ errors. The Hamming distance for the Mariner telemetry code is 16, so up to seven errors can be detected and corrected.

All this still leaves open precisely how each pixel is encoded as one of the rows of M_5. Certainly, the gray level representing the pixel may be encoded itself as a five-digit binary number. It would be tempting simply to transmit that pattern of 0s and 1s, but errors would play havoc with the information; it would be impossible to ever know what the bits originally transmitted were.

So it is that the gray level x, when it is generated on board as a binary number

$$(x_4, x_3, x_2, x_1, x_0)$$

is encoded as the kth row of M_5, where k is the binary number with these bits. At the ground station, as already implied, each received word is matched against the rows of M_5, the best match selected, and the corresponding 5-bit vector generated and output to a tape file. When the signals marking the end of a picture are received, the tape is played out through graphic display units for visual inspection.

The decoding process is really much more complicated than this because the message rate of the spacecraft is over 16,000 bits/s. The ground computer must therefore work very quickly indeed to transform this information to the resulting 500 pixels per second. In fact, the burden of this work was carried out in the later Mariner missions by a special-purpose computer embodying a discrete version of the fast Fourier transform (see Chapter 32).

The previous code is called *error-correcting* because it permits the receiver of a message so encoded to correct (up to a certain number of) errors that may have entered the message during its long flight through space. An *error-detecting code* is one that enables the receiver to detect (up to a certain number of) errors. Both error detection and error correction may be readily understood in terms of the Hamming distance. If the Hamming distance between a transmitted word and the same word when received is h, then h errors have occurred. If the Hamming distance between any two code words (as transmitted) is always d or more, however, then h can be as large as $d-1$ and the receiver will still know that errors have occurred. If exactly d errors occur, there is a possibility that one code word has merely been corrupted into another; the receiver is none the wiser.

If errors are thus detected, what does the receiver do? The receiver simply selects the code word in the list of code words that is closest to the received word in terms of the Hamming distance. The choice will obviously be correct if no more than $\lfloor(d-1)/2\rfloor$ errors occurred.

Problems

1. How many 6-bit words can you find that are all at mutual Hamming distances of 3 or more from each other? How many errors will such a code detect? Correct?

2. Develop the (16, 4) Reed-Muller code by deriving M_4. What are its error-detection and correction capabilities?

References

E. C. Posner. Combinatorial structures in planetary reconnaissance. *Error Correcting Codes* (H. B. Mann, ed.). Wiley, New York, 1969.

Richard W. Hamming. *Coding and Information Theory.* Prentice-Hall, Englewood Cliffs, N.J., 1980.

BOOLEAN LOGIC

Expressions and Circuits

At the heart of every computer there is a logic control unit which orchestrates the transfer and manipulation of information between and within thousands of sites where information is stored. Essentially, this unit is an electronic circuit with dozens of input and output lines. It passes incoming information through a network of logic gates (see Chapter 48), and the particular configuration of gates used determines precisely the function which the logic control unit will perform. To put it the other way round, first computer designers decide what function or functions they want the unit to carry out, and then they design the circuitry to do just that. Of course, logic circuitry is used in a great many devices besides computers.

The progression from the functional specification of a logic circuit to the circuit itself generally involves an intermediate step in which the function is expressed as a formula of one kind or another. This three-step process embraces a hundred year period from the time of George Boole in the middle nineteenth century to the present. It was Boole who provided much of the intellectual machinery for such formulas and their manipulation.

A *boolean variable* can have one of only two possible values, namely, 0 or 1. A

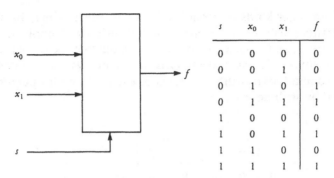

s	x_0	x_1	f
0	0	0	0
0	0	1	0
0	1	0	1
0	1	1	1
1	0	0	0
1	0	1	1
1	1	0	0
1	1	1	1

Figure 13.1 The multiplexer function

boolean function has a number of input variables, all boolean, and for each possible combination of input values, such a function has a boolean-valued output, either 0 or 1. Figure 13.1 shows a device and a table which specifies the function it embodies. For now the interior of the device is shown empty. Later we will fill it in.

Figure 13.1 specifies the simplest possible form of "multiplexer", whose purpose is to combine this information carried on the two lines x_0 and x_1 into a single line f. Specifically, it allows the information on one of the lines at a time to pass through its circuitry and out onto line f. Which line will have this privilege at any time is decided by a third input line labeled s, for "select".

When the s input has value 0, the value currently in line x_0 is transmitted along f; and when s equals 1, the value in x_1 is so transmitted. This functional description is completely summarized in the truth table in which, for each of the eight possible combinations of boolean input values, a corresponding output value (labeled f) is listed. In this particular case, we see that in every row where $s = 0$, f has the same value as x_0 and in every row where $s = 1$, $f = x_1$.

The function f just described has three variables, and its truth table (so named because 0 and 1 are sometimes referred to, respectively, as *false* and *true*) has eight rows. The particular pattern of 0s and 1s in the f column serves to specify f completely; a different pattern means that a different function is being specified, and there are, all told, $2^8 = 256$ such patterns, that is, 256 three-variable boolean functions are possible. Now the truth table for a four-variable boolean function has 16 rows, and so there are $2^{16} = 65,536$ possible four-variable functions. Clearly, the number of boolean functions grows pretty quickly as n increases. In fact, for n-variable boolean functions, this number is obviously

$$2^{2^n}$$

Just as with other kinds of variables, boolean variables may be added and multiplied, after a fashion, and even negated. By using such boolean operations (denoted by $+$, $\cdot$, and $'$, respectively), it is possible to build up expressions of great complexity. Given two boolean variables x and y, we may summarize each of the three operations by truth tables, just as each arithmetic operation may be summarized by arithmetic tables.

x	y	$+$
0	0	0
0	1	1
1	0	1
1	1	1

x	y	$\cdot$
0	0	0
0	1	0
1	0	0
1	1	1

x	$'$
0	1
1	0

Another way of looking at these operations involves the more traditional view of 0 as false and 1 as true. In this setting, we may think of $+$ as OR, $\cdot$ as AND, and $'$ as NOT. So $x + y$ means, in effect "x or y," and such a statement is true if at least one of x or y is true. Thus $x + y = 1$ if at least one of x OR y equals 1.

Astute readers will have noticed that $+$ and $\cdot$ are just two of $2^{2^2} = 16$ possible two-variable boolean functions. Might not other such functions serve equally well as boolean operations? Indeed, they do; some actually correspond more closely to the actual gates used in modern computer logic circuits (see Chapter 3).

As far as building up complex *boolean expressions* is concerned, the process is really quite simple. Assume we have a set Σ of symbols which will be used to denote boolean variables.

If $x \in \Sigma$, then x is a boolean expression.
If A and B are boolean expressions, then so are $(A) + (B)$, $(A) \cdot (B)$, and $(A)'$.
Nothing else is a boolean expression.

These three rules enable us to start with variables like x, y, z and end up with expressions like

$$(((x + (y'))') \cdot (z \cdot (y \cdot (x'))))$$

Because of all the parentheses, expressions like this make for cluttered reading. But in practice most of the parentheses can be stripped away, either because of operator precedence rules or because of algebraic simplifications. For example, we give $\cdot$ precedence over $+$ so that an expression like $x + y \cdot z$ means $x + (y \cdot z)$ and not $(x + y) \cdot z$. Similarly, $'$ has precedence over $+$ or $\cdot$, so that $x + y'$ means $x + (y')$ and not $(x + y)'$. By application of such rules, the expression above can be simplified to

$$(x + y')' \cdot (z \cdot (y \cdot x'))$$

The fact that $z \cdot (y \cdot (x'))$ can be written more simply as $z \cdot y \cdot x'$ follows from the associativity of the $\cdot$ operation which in turn follows from the axioms of boolean algebra, first formulated by Boole in 1854.

This algebra in its modern form is defined as follows: B is a set, $+$ and $\cdot$ are regarded as two binary operators on the set, and $'$ is a unary operator. Recall that $\forall$ means "for all" and that $\exists$ means "there exists."

1. $x + y \in B$ and $x \cdot y \in B$, $\forall x, y \in B$
2. $\exists 0$ and $1 \in B$ such that $x + 0 = x$ and $x \cdot 1 = x$, $\forall x \in B$
3. $x + y = y + x$ and $x \cdot y = y \cdot x$ $\forall x, y \in B$
4. $x \cdot (y + z) = (x \cdot y) + (x \cdot z)$ and $x + (y \cdot z) = (x + y) \cdot (x + z)$, $\forall x, y, z \in B$
5. $\forall x \in B$ $\exists x' \in B$ such that $x + x' = 1$ and $x \cdot x' = 0$

These axioms sum up everything which Boole meant to express by the logic connectives like AND, OR, and NOT. In such a context, it is not terribly difficult to see that all the axioms are true under the ordinary logic interpretation. However, axioms are intended as a starting point for deductions. As such, they comprise the foundation of a system, and they themselves are taken as primary.

For example, from the axioms above, we can deduce the associative law for multiplication:

$$x \cdot (y \cdot z) = (x \cdot y) \cdot z$$

and this enables us finally to simplify the expression

$$(x + y')' \cdot (z \cdot (y \cdot x'))$$

to

$$(x + y')' \cdot (z \cdot y \cdot x')$$

and to

$$(x + y')' \cdot z \cdot y \cdot x'$$

Generally speaking, we may use the many algebraic tools deduced from the axioms above to manipulate boolean expressions, especially with a view to simplifying them further. We will see an example of such a simplification process shortly.

In the meantime we note something very significant about boolean expressions as we have defined them: Each boolean expression defines a boolean function because such an expression can be evaluated for each combination of values for its variables. For example, when $x = 0$, $y = 1$, and $z = 1$, in the expression above we get

$$(x + y')' \cdot (z \cdot (y \cdot x')) = (0 + 1')' \cdot (1 \cdot (1 \cdot 0'))$$
$$= (0 + 0)' \cdot (1 \cdot 1)$$
$$= 0' \cdot 1$$
$$= 1 \cdot 1$$
$$= 1$$

Even more significantly, we can turn this statement around: Each boolean function is defined by some boolean expression (in fact, by a whole infinity of such expressions). The simplest way to see that is to examine the truth table of a function for which we require a boolean expression:

x	y	z	f
0	0	0	0
0	0	1	0
0	1	0	0
0	1	1	1
1	0	0	1
1	0	1	0
1	1	0	1
1	1	1	1

For each row at which $f = 1$, examine the corresponding values of x, y, and z. Write a product in which the variable appears negated if it has value 0 and in which it appears unnegated if it has value 1. Thus the row 011 above yields the product

$$x' \cdot y \cdot z$$

Now add all the products obtained in this way:

$$x' \cdot y \cdot z + x \cdot y' \cdot z' + x \cdot y \cdot z' + x \cdot y \cdot z$$

and notice two things:

The only time the resulting function can have the value 1 is when one of the products has that value.

The only time one of the products can equal 1 is when its constituent variables have the values in the row (of the truth table) corresponding to that product.

The expression above therefore realizes the function f, and such expressions have a special name — *disjunctive normal form.* Not surprisingly, every function can also be written as a product of sums, in *conjunctive normal form.*

If we substitute x_0 for x, x_1 for y, and s for z in the previous truth table, we recognize the multiplexer function defined at the beginning of this chapter. So we now have in hand a boolean expression which describes the multiplexer precisely:

$$f = x_0' \cdot x_1 \cdot s + x_0 \cdot x_1' \cdot s' + x_0 \cdot x_1 \cdot s' + x_0 \cdot x_1 \cdot s$$

The second step of the process described in this chapter is to move from a boolean expression to a logic circuit. However, the logic circuit which corresponds to the last expression is fairly complicated, and generally speaking, the shorter the expression, the simpler the circuit. If we are in the business of building actual hardware, then we want all our circuits (and, therefore, our expressions) to be as simple as possible.

One technique for simplifying boolean expressions is to exploit the axioms and theorems of boolean algebra:

$$
\begin{aligned}
f &= x_0' \cdot x_1 \cdot s + x_0 \cdot x_1 \cdot s + x_0 \cdot x_1' \cdot s' + x_0 \cdot x_1 \cdot s' \qquad \text{by axiom 3}\\
&= x_1 \cdot s \cdot (x_0' + x_0) + x_0 \cdot s' \cdot (x_1' + x_1) \qquad \text{by axioms 3 and 4}\\
&= x_1 \cdot s \cdot 1 + x_0 \cdot s' \cdot 1 \qquad \text{by axiom 5}\\
&= x_1 \cdot s + x_0 \cdot s' \qquad \text{by axiom 2}
\end{aligned}
$$

Another technique for simplifying expressions involves the use of special tables (see Chapter 20).

Each boolean expression corresponds to a unique logic circuit, and the simplest way to describe this correspondence formally is to imitate the definition of a boolean expression. As the expression is constructed, so is the circuit:

A single line labeled x is a logic circuit. One end is regarded as the input, and the other end is output.

If A and B are logic circuits with output lines a and b, then so are the circuits shown in Figure 13.2.

Nothing else is a boolean circuit.

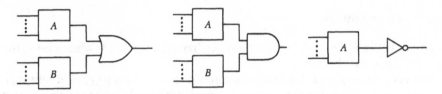

Figure 13.2   Steps in constructing a boolean circuit formally

Three gates are shown in Figure 13.2. The OR gate (shaped like a shield) computes $A + B$: Its output is a 1 if and only if at least one of A or B is inputting a 1 to the gate. The AND gate (shaped like a semicircle) computes $A \cdot B$, outputting a 1 if and only if both A and B input a 1 to the gate. The inverter merely computes A' by changing a 1 to a 0 and conversely.

Conventions for simplifying the representation of circuits are similar to those used in stripping away the redundant parentheses in a boolean expression. This enables us to have gates with more than two inputs, for example, and results in the circuit shown in Figure 13.3 for the disjunctive normal form of the multiplexer expression.

The complexity of this circuit is obvious at a glance. Measured objectively, the circuit has 8 gates and 25 lines. Compare this circuit with the one shown in Figure 13.4, which corresponds to the simplified version of the multiplexer's boolean expression.

This circuit has 4 gates and 8 lines and is clearly much simpler—yet it

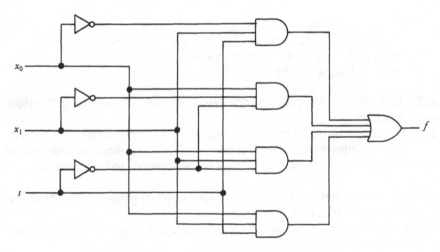

Figure 13.3   A complicated multiplexer circuit

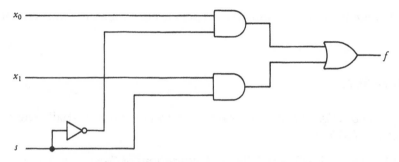

Figure 13.4 A simple multiplexer circuit

computes the same function! Indeed, we would certainly rather put this circuit inside the multiplexer box, shown earlier, than the previous circuit.

It would not be true to say that modern computers use AND, OR, and NOT gates directly. In fact, the field-effect transistors employed in modern silicon technology (see Chapter 56) correspond to gate types different from these, for example, the NAND gate (see Chapter 3) which, for two inputs x and y, computes $(x \cdot y)'$, that is, NOT (x AND y).

The multiplexer we have just designed will be used later in the design of a simple computer (see Chapter 48).

Problems

1. The conjunctive normal form for a boolean function f on three variables consists of a product of sums, with each sum containing three literals. How would you use the truth table of f to obtain an expression for f in this form?

2. Using the axioms for boolean algebra, prove the validity of the so-called De Morgan's laws:

a. $(x + y)' = x' \cdot y'$

b. $(x \cdot y)' = x' + y'$

3. Obtain a conjunctive normal form for the multiplexer, and draw the corresponding circuit. How does it compare, in terms of number of gates, with the circuit in Figure 13.3?

4. A "demultiplexer" has a function reverse to that of a multiplexer. A single input enters the device, and two (say) inputs leave it. A select line determines which route incoming information takes in leaving the demultiplexer. Design one.

References

M. M. Mano. *Digital Logic and Computer Design.* Prentice-Hall, Englewood Cliffs, N.J., 1979.

Alfred V. Aho and Jeffrey D. Ullman, *Foundations of Computer Science,* Chapter 12. W. H. Freeman, New York, 1992.

REGULAR LANGUAGES

Pumping Words

Besides programming languages, there are other, more abstract languages studied by computer scientists. Their names hint mildly at their natures (also see Chapter 7):

Regular languages
Context-free languages
Context-sensitive languages
Recursively enumerable languages

In the classification of abstract languages, it is frequently useful to know when a given language is *not* in a particular class. One technique for doing precisely this is based on the *pumping lemma,* first published in 1961 by Y. Bar-Hillel, M. Perles, and E. Shamir. The pumping lemma for regular languages tells us that if we select a sufficiently long word from a regular language and "pump" it—as many times as we like—we always get a new word in that language (Figure 14.1).

What makes such knowledge useful is that sometimes, in dealing with a word

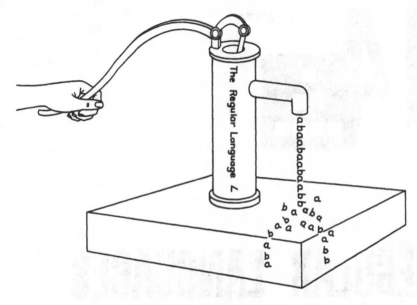

Figure 14.1 The language pump

from a language we know little about, we may find that a particular word in it cannot be pumped in any way without getting a word *not* in the language. In such a case, we immediately know that the language was not regular.

To *pump* a word *W* means to select a particular subword, say *Y*, from *W* and to replace *Y* by *YY* in *W*. Symbolically, if

$$W = XYZ$$

then the result of pumping *W* (at *Y*) once is

$$W' = XYYZ$$

and the result of pumping *W* twice is

$$W'' = XYYYZ$$

and so on. Indeed, we can pump *W* "negatively" by eliminating *Y* altogether from *W*.

It is possible to discover both what the pumping lemma says and how it works, simultaneously, by examining a portion (Figure 14.2) of the state-transi-

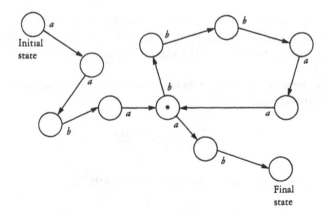

Figure 14.2 An input word leaves a trail

tion diagram for the finite automaton A which accepts a particular regular language L (see Chapter 2). Here different states of the automaton are represented by circles, and transitions between states are represented by arrows labeled with the input symbol which causes that particular transition.

Each word in L corresponds to a trail through A's diagram. The trail leads from the initial state to one of A's final states. For example, the trail for the word *aababbbaaab* is shown in the figure above. The trail is simply the sequence of transitions through which A must go in accepting the word.

The most interesting thing about the trail pictured above is that it contains a loop: having entered one of A's states, the trail returns to it later. This observation gives us an immediate idea for producing new words which A must accept: Simply repeat that portion of the word corresponding to the loop

$$aaba\,(bbbaa)\,ab$$

to get

$$aaba\,(bbbaa)\,(bbbaa)\,ab$$

In other words, *aababbbaabbbaaab* must also be accepted by A and hence must be in L. By the same reasoning, we can go around the loop as many times as we like before going on to the final state. Each time, we get a new word in L.

Unfortunately, there is nothing to guarantee that the trail corresponding to an arbitrarily selected word in L has a loop! However, A has only a finite number of states, say n of them, and a word of length $\geq n$ which happens to be in L will certainly drive A through just such a loop: any word of length n or more must

93

cause at least one state to be revisited. This simple observation is an example of the *pigeonhole principle*. If more than n pieces of mail are delivered to n pigeonholes, then at least one of the pigeonholes must receive more than one piece of mail.

One more observation will put us in a position to state the pumping lemma. We note that the subword

$$XY = \underbrace{aaba}_{X}\ \underbrace{bbbaa}_{Y}$$

has total length $\leq n$ while the subword Y has length ≥ 1.

The Pumping Lemma for Regular Languages

If L is a regular language, then there is a constant n such that for each word W in L having length $\geq n$, there are words X, Y, Z such that

$W = XYZ$
Length of $XY \leq n$
Length of $Y \geq 1$
XY^kZ is in L for $k = 1, 2, 3, \ldots$

Here, Y^k simply means k repetitions of the word Y strung together in the fashion we are already familiar with.

It is enjoyable to use the pumping lemma to show that a particular language is *not* regular. For example, let L be the language consisting of all palindromes over the alphabet $[a, b]$. Such words are symmetric about their midpoints, for example,

abbababba

If L *is* regular, then the pumping lemma applies to L and there would be some constant n (unknown, in general) which we must use to measure the length of candidate words for the pumping lemma's application. The strategy to use here is to notice that no matter what value n might have, the word

$$W = a^nba^n$$

must be in L according to the definition of palindromes. But, according to the pumping lemma, W can be written

$$W = XYZ$$

so that, in particular, the length of XY, the initial part of W, is less than or equal to n. Thus, XY consists of nothing but a's, and Y consists of at least one a. It now follows, by the pumping lemma, that the word

$$W = XY^2Z = a^mba^n$$

is also in L. Now a^m reflects the fact that some nonzero portion of the initial string a^n has been pumped, so that $m > n$. But in this case, $a^m ba^n$ cannot possibly be in L since it is *not* a palindrome! The only thing that can have gone wrong in all this reasoning is the initial assumption that L was regular. Evidently, it is not.

There are pumping lemmas (not exactly like the one for regular languages but similar in spirit) for other sorts of abstract languages as well, including context-free languages.

Problems

1. Since finite automata cannot recognize palindromes, it comes as no great surprise to learn that they cannot recognize prime numbers either. If we represent a prime number p by a string of p ones, we might ask whether we could find a finite automation which accepted all such strings (and, naturally, *only* such strings). Use the pumping lemma to show that no such finite automation can exist.

2. Prove the pumping lemma by formalizing and amplifying the ideas preceding its statement. In particular, assume a finite automaton with n states and alphabet $\Sigma = \{a, b\}$.

References

J. E. Hopcroft and J. D. Ullman. *Introduction to Automata Theory, Languages and Computation.* Addison-Wesley, Reading, Mass., 1979.

Daniel I. A. Cohen. *Introduction to Computer Theory.* John Wiley, New York, 1986.

TIME AND SPACE COMPLEXITY

The Big-O Notation

When a program is run on a computer, two of the most important considerations are how long it will take and how much memory it will use. There are other issues such as whether the program will even work (see Chapter 10), but the two concerns of computation time and memory space dominate our thinking in a number of computing environments. For example, on large mainframe computers, user charges are calculated on the basis of the time a program has run and how much memory it used. Even on small computers, we want a program to run quickly and not to exceed the amount of available memory.

There is a very simple problem that can be used to illustrate these two aspects of algorithmic efficiency. Suppose *n* positive integers are stored in array *A*. Are the integers all distinct, or are at least two of them the same? It is not always easy to spot duplicate integers by eye:

A: 86, 63, 39, 98, 96, 38, 68, 88, 36, 83, 17, 33, 69, 66, 89, 96, 93

A straightforward algorithm for detecting duplicates considers one integer at a time and scans array A, looking for a match:

SCAN

for $i \leftarrow 1$ **to** $n - 1$ **do**
  for $j \leftarrow i + 1$ **to** n **do**
    if $A(i) = A(j)$
      then output i and j;
        exit
      else continue

The algorithm SCAN uses very little storage, just the n array locations and the two index variables i and j. Consequently, we could say that SCAN requires $n + 2$ words of storage, assuming that each word is large enough to contain any integer we wish.

To determine how much time SCAN requires, we use a simple charting technique which assumes that each line of the program costs 1 unit of time to execute. To detect a duplicate among four positive integers SCAN's execution pattern may be charted as follows:

for $i \leftarrow 1$ **to** $n - 1$ **do**
  for $j \leftarrow i + 1$ **to** n **do**
    if $A(i) = A(j)$
      then output i and j;
        exit
      else continue

Each dot on the chart represents one execution of the instruction occupying the same level as the dot. Lines connect the dots in a top-to-bottom, left-to-right order as a representation of how the algorithm executes for a specific instance of the problem. In the example above, the last vertical sequence of dots represents the point where $i = 3$ and $j = 4$. In this case the test "**if** $A(3) = A(4)$" passed, and the algorithm halted after outputting 3 and 4 before halting.

Since the chart above has 22 dots, we could say that SCAN requires 22 units of time to process the sequence 7,8,4,4. If the sequence had been 7,8,4,5 then the last vertical vow of dots would have looked like the others. At least we shall adopt this convention; having reached the **continue** statement after transferring around the **output** and **exit** statements, the algorithm would have discovered it was out of indices and quit anyway.

This example raises some very interesting questions. First, given n integers, what is the greatest number of time units that SCAN will require to determine whether the sequence contains duplicate integers? Another question of some importance is, How long, on average, will SCAN require to make this determination? We tackle only the first question here.

Evidently, SCAN always takes the longest time on a sequence of n integers when the last two integers (and only those two) are duplicates. First, each execution of the inner loop when the **if** test fails requires 3 steps. When SCAN starts with $i = 1$, j runs from 2 to n and the inner loop is executed $n - 1$ times for a total of $3(n - 1)$ steps. This yields $1 + 3(n - 1)$ steps for all instructions executed while $i = 1$. When $i = 2$, it is not hard to see that SCAN will execute $1 + 3(n - 2)$ steps. The expression for the total time taken by SCAN is

$$1 + 3(n - 1) + 1 + 3(n - 2) + \cdots + 1 + 3(2) + 1 + 4$$

$$= n + 1 + 3 \sum_{k=1}^{n-1} k = n + 1 + \frac{3n(n-1)}{2} = \frac{3n^2 - n + 2}{2}$$

The *worst-case time complexity* of an algorithm operating on an input of size n is simply the maximum time that the algorithm will spend on any input instance of size n. The worst-case complexity of SCAN, by the measure we have adopted, is $(3n^2 - n + 2)/2$. Of course, this formula is based on the convenient fiction that all instructions in a program absorb the same amount of time. Indeed, this is not the case in real programming languages running on real computers. But it is frequently possible to assign actual execution times (microseconds or less) to individual instructions and to repeat an analysis essentially like that given above. In such a case, one might obtain a formula with different coefficients, say $(7.25n^2 - 1.14n + 2.83)/2$. Any conceivable embodiment of SCAN would yield a worst-case running-time formula that is quadratic.

For this reason, we have agreed to use an order-of-magnitude notation when expressing time and space complexity for either algorithms or programs. A function $f(n)$ is said to be the *big O of a function $g(n)$* if there is an integer N and a constant C such that

$$f(n) \leq c \cdot g(n) \qquad \text{for all } n \geq N$$

We write

$$f(n) = O(g(n))$$

And if it should happen that $g(n) = O(f(n))$ as well, then the two functions are said to have the same order of magnitude.

Turning now to SCAN and denoting its worst-case time complexity by $T_w(n)$, we have

$$T_w(n) = O(n^2)$$

It turns out that all quadratic functions of n have the same order of magnitude. For this reason, it makes sense to write the simplest quadratic function of n available to indicate the order-of-magnitude time complexity of SCAN.

Clearly, SCAN requires only $O(n)$ storage space since $n + 2 = O(n)$.

As a contrast to SCAN, we now examine a different algorithmic approach to the problem. The algorithm STOR adopts the simple tactic of storing each integer in array A in another array B at an index equal to the integer itself!

STOR

```
for i←1 to n do
    if B(A(i)) ≠ 0
        then output A(i);
            exit
        else B(A(i)) ←1
```

It is assumed that array B initially contains 0s. Each time an integer a is encountered, $B(a)$ is set to 1. In this way previously encountered integers are detected by the **if** statement.

The use of a number to be stored as the basis for a computation of its storage address underlies the technique known as *hashing* (see Chapter 43). In the primitive version of hashing used above, we must assume that array B is large enough to contain any of the integers stored in A.

It is instructive to compare the worst-case time complexity of STOR with that of SCAN. Consider, for example, its performance on the same four-number sequence 7,8,4,4:

```
for i←1 to n do
    if B(A(i)) ≠ 0
        then output A(i);
            exit
        else B(A(i)) ←1
```

The time complexity of STOR on this sequence is 13 whereas SCAN required 22 steps. This happens to be the worst case for STOR when $n = 4$, and in general STOR's worst-case complexity is

$$T_w(n) = 3n + 1 = O(n)$$

On the other hand, STOR requires far more storage than SCAN: If numbers of up to m bits are involved, then STOR requires at most 2^m memory locations for correct operation.

By comparing the worst-case time complexity of SCAN and STOR, it becomes immediately obvious which algorithm is superior when the two complexities are plotted (Figure 15.1).

Although SCAN takes less time than STOR on sequences of lengths 1 and 2, by the time $n = 3$, STOR is already showing its superiority. Such a conclusion would be drawn no matter what time complexities the two algorithms had—as long as SCAN was quadratic and STOR was linear. Sooner or later an algorithm with linear time complexity will outperform one with quadratic complexity.

The same thing is true of the space complexity of algorithms: An algorithm with quadratic memory requirements will generally exceed the storage capacity of a computer much sooner than an algorithm with linear space complexity.

Studies of the time and space complexity of algorithms tend to concentrate on the question of how fast a given problem can be solved algorithmically. Unless storage requirements are exorbitant, the main question about a given algorithm's efficiency is usually, Can we find a faster algorithm to solve the same problem?

Suppose there is a problem P for which the fastest known algorithm has a

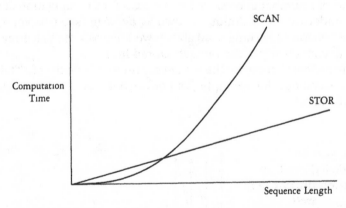

Figure 15.1 The worst case time complexities of SCAN and STOR

worst-case complexity $O(f(n))$. If it can be shown that no faster algorithm exists, one which has complexity $g(n)$ where $g(n) = O(f(n))$ but $f(n) \neq O(g(n))$, then the problem itself can be said to have complexity $O(f(n))$. For example, under reasonable assumptions about how a sorting algorithm may make comparisons, it can be shown that sorting n integers is an $O(n \log n)$ problem (see Chapter 40). Unfortunately, very few other problems have known complexity; proving that a given algorithm is the fastest possible for a given problem is notoriously difficult.

Indeed, there are some problems which appear to have exponential complexity (see Chapter 41). No one has ever found a polynomial-time algorithm for any of them. Although an $O(2^n)$ algorithm may exist, no one has discovered an algorithm which runs in time which is even $O(n^{1000})$. It is not hard to show that when n is large enough, $n^{1000} < 2^n$.

All problems that we might conceivably wish to solve and that have a reasonable measure n of instance size may presumably be ranked in order of complexity. At the top of the list come those problems requiring $O(2^n)$ steps (or more) to solve. They are forever out of practical reach for a sequential computer. Next come those problems having polynomial-time solutions — in order of the polynomials. Some problems may have complexity $O(n^2)$; others may have complexity of $O(n)$ or even less. In fact, there are problems of intermediate complexity, for example,

$$O(n) < O(n \log n) < O(n^{1/2}) < O(n^2)$$

The combinatorial mathematician Jack Edmonds was the first to draw attention forcibly to the distinction between exponential- and polynomial-time problems and their algorithms in 1965. It seemed very strange that some problems, such as finding the shortest path between two points in a graph, should have an $O(n^2)$-time algorithmic solution while other, seemingly closely related problems, such as finding the longest such path, should have only an $O(2^n)$-time algorithmic solution. This distinction was to be explored and, to an extent, explained by Stephen Cook in 1971 (see Chapter 45).

Problems

1. The STOR algorithm operates correctly only if it uses an array A that is already 0-filled. Write a new algorithm that has the same time complexity as STOR but does not require A to be 0-filled.

   (*Hint:* Use an auxiliary array B having at least n entries.)

2. Show that if $\epsilon > 0$. then

$$\log n = O(n^\epsilon)$$

but

$$n^\epsilon \neq O(\log n)$$

3. Given that a sequence of n integers contains exactly two duplicates, what is the *average time complexity* of STOR? Over all such sequences, how long does STOR take (to detect the duplicates) as a function of n?

4. The way in which the size of a problem is measured sometimes makes a big difference in the time complexity of an algorithm for that problem. For example, in the problem solved by SCAN, each integer in the input sequence counted once toward n, the problem size. However, given a single integer m as input, we may count it as *(a)* size 1, *(b)* size *log m*, or *(c)* size *m*. Devise a straightforward algorithm for determining whether m is prime. Show that under (a) there is no upper bound on how long your algorithm will take. Show that under *(b)* your algorithm has time complexity at least $O(2^m)$, whereas under *(c)* it has polynomial-time complexity.

References

Alfred V. Aho and Jeffrey D. Ullman. *Foundations of Computer Science,* Chapter 3. W. H. Freeman, New York, 1992.

Daniel H. Greene and Donald E. Knuth. *Mathematics for the Analysis of Algorithms.* Birkhauser, Boston, 1982.

16

GENETIC ALGORITHMS

Solutions That Evolve

If organisms as complicated as plants and animals can evolve in a primordial soup, why can't solutions to a problem evolve in a computer's memory? The short answer is "They can." At least they can if they're given enough time: not eons, but anywhere from seconds to hours.

The theory of natural selection states that variability in a natural population (due to mutations and new, sexually produced gene combinations) helps to ensure the population's survival and fitness. When the environment changes, variability in the population may mean that some individuals are better adapted to the new conditions than others. Such individuals form the basis for a new, more fit population.

The same principle applies to genetic algorithms. A simple problem illustrates the idea in a computer setting. Suppose we want to find the maximum value of a one-variable function, F, with an independent variable, x, that lies between 0 and 1. Somewhere in this interval there is a point x_m at which F achieves its maximum value. It will be the job of the genetic algorithm to find x_m to within, say, 1/64th of its true value.

Figure 16.1 illustrates the problem in the case of a two-variable function. Its

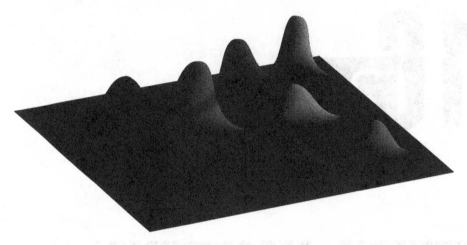

Figure 16.1 A functional landscape

values, when plotted, may produce a landscape of peaks and valleys. A genetic algorithm that finds the optimum of the one-variable case could certainly be adapted to solve the two-variable case.

As far as the one-dimensional case is concerned, somewhere in the interval [0, 1] lies the potential solution that must be expressible in some genetic form. Since we are interested only in solutions that are accurate to six bits, the algorithm will use a *chromosome* consisting of six genes, one for each bit. The chromosome may be written *abcdef* and the particular number represented by the chromosome will depend on the *alleles,* or values, of the bits. Thus the number 23/64 will be .010111 and the alleles for this particular chromosome will be $a = 0$, $b = 1$, $c = 0$, $d = 1$, $e = 1$, and $f = 1$.

Genetic terminology aside, the algorithm proceeds by setting up an initial population of solutions created at random in this particular example. The population need not be very large in relation to the total space of possibilities, say just 10 chromosomes.

No animals are represented by these chromosomes but solutions are. The *fitness* of each solution is evaluated by simply calculating the value of the function at each of the points represented by the chromosomes. That is exactly what the algorithm does in its initial step. Then it creates new individuals by a mixture of mutation and crossover. An individual chromosome mutates when one of its genes switches its allele from one value to the other. For example, 011010 is a mutation of 011110; a 1 changed to a 0. The crossover operation requires two chromosomes from the current population, just as a new animal in a population requires two parents. The chromosomes are crossed at some gene. This means that if the chromosomes are

110010 and **010001,**

and if they are crossed over between the third and fourth gene, then one of the offspring will be 110**001**. The other offspring will be **010**010.

In this manner, a genetic algorithm generates new, potentially better solutions from old ones. There are numerous strategies, obviously, for selecting solutions from the current population which will form the basis of the next generation. Almost all strategies involve a mixture of crossover and mutation. For example, the present algorithm

a) selects the top six solutions and breeds them in three pairs to obtain three new "progeny."
b) mutates one of the top six solutions and one of the bottom four.

The algorithm next evaluates the "fitness" of the resulting 15 solutions by evaluating F at each of them, retaining the solutions with the highest 10 values and discarding the rest.

A genetic algorithm will thus churn its way through generation after generation of potential solutions. As a general rule the fitness of the population continues to increase up to some point identified as a *steady state* value. No more improvement can be found once the population has reached this seeming plateau. I say "seeming" because in some problems, it is perfectly possible for the fitness measure to get stuck for long periods of time at a single level.

How close is the steady state value to the value of x at which F reaches its true maximum? If one happens to know the function F one may immediately compare x_m with the best solution x_b that the algorithm was able to find. If $|x_b - x_m| < \frac{1}{64}$, the algorithm succeeded.

If one does not happen to know the true value of x_m, what sort of reliance can be placed on x_b, the solution found by the genetic algorithm? The answer in this case, as in all applications of genetic algorithms, depends on the functional properties of the problem (F, in this case). If F has a single maximum and is reasonably well behaved showing a suitable amount of variation, the genetic algorithm will produce good estimates of x_m is perhaps 10 to 20 population cycles.

If the problem function F has multiple maxima, the genetic algorithm may take much longer to find the true maximum.

Genetic Algorithms, pioneered by John H. Holland at the University of Michigan in the 1960s, have been applied to a great many problems, including the classical problems of algorithmic analysis such as the traveling salesman problem.

Problems more complicated than the simple function maximization problem may require more elaborate schemes to encode problem instances genetically. How, for example, should one encode potential solutions for the traveling salesman problem? (See Chapter 41.)

Suppose that a traveling salesman has six cities to visit, A, B, C, D, E, and F. The cost of a round-trip ticket to the six cities will vary, depending on the route taken. Thus to visit the cities in the order BDAFCE may be cheaper than to visit them in the order ABCDEF. Unfortunately, the sequences themselves cannot be used as chromosomes since the crossover operation applied to two of them may not even result in a tour.

A clever method of getting around this difficulty uses a standard ordering, say ABCDEF, giving it the code 123456. All other sequences have a code that is based on removal of letters from the standard order. A sequence like BDAFCE, for example, has the code 231311. The procedure is simple: remove each letter of the second sequence one at a time from the standard sequence. When a letter is removed, note its position in the standard sequence at the time of its removal.

B is 2 in ABCDEF
D is 3 in ACDEF
A is 1 in ACEF
F is 3 in CEF
C is 1 in CE
E is 1 in E

Thus, BDAFCE has the chromosome 231311. Interestingly enough, any sequence of six digits will turn out to be a tour provided that the ith digit does not exceed $7 - i$; $i = 1, 2, \ldots 6$. (Mathematically inclined readers may take a moment aside to prove this for themselves.) When two such sequences are spliced together, say 231311 and 512121 with a crossover between the fourth and fifth positions, the resulting codes, 231321 and 512111, turn out to represent tours as well: BDAFEC and FACBDE (Figure 16.2).

In the application of genetic algorithms, considerable ingenuity must sometimes be applied to the *representation problem*. How does one encode proposed solution instances into a chromosome which can not only be manipulated meaningfully, but also allows rapid evolution of the true or optimum solution?

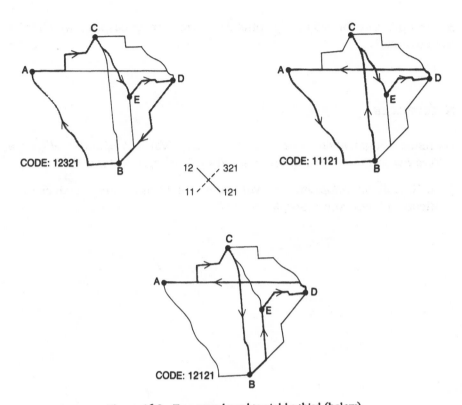

Figure 16.2 Two tours breed to yield a third (below)

Problems

1. Using the methods of Chapter 15, evaluate the computational complexity of the population cycle for the maximum-finding genetic algorithm described in this chapter.

2. Assuming that $F = 8x^2 + 8x - 1$, analyze what happens when two solutions are crossed. How often does the offspring show an improvement over either parent? How often does a random mutation produce a superior solution? Can you develop a "theory of genetic algorithms" for this function, based on the sample algorithm?

3. Is it possible for two nearly optimal tours to have an offspring which is far from optimal?

References

Lawrence J. Fogel, Alvin J. Owens, and Michael J. Walsh. *Artificial Intelligence Through Simulated Evolution.* John Wiley & Sons, New York, 1966.

John H. Holland. *Adaptation in Natural and Artificial Systems.* University of Michigan Press, Ann Arbor, Mich., 1975.

17

THE RANDOM ACCESS MACHINE

An Abstract Computer

The term *random access machine (RAM)* tends to have a double use among computer scientists. Sometimes it refers to specific computers with *random access memories,* that is, memories in which the access to one address is just as fast and easy as access to any other address. At other times, it refers to a certain model of computation which is no less "abstract" than, say, a Turing machine, but which is closer in its operation to standard, programmable digital computers.

Not surprisingly, the main difference between Turing machines and (abstract) RAMs lies in the kind of memory employed. On one hand, a Turing machine's memory is all on tape, making it a sort of "serial access" machine. A RAM, on the other hand, has its memory organized into words, with each word having an address. It has, moreover, a number of registers. The model described here has one register called the *accumulator,* (AC for short).

As was the case with Turing machines, RAMs are actually defined in terms of programs, the sort of structure shown in Figure 17.1 being merely a handy vehicle for interpreting such programs. As such, we picture a RAM as being equipped with a control unit able to carry out all the operations specified by a RAM program with which it is "loaded" in some sense. The program dictates the transfer of information between various memory words and the AC. It also specifies certain operations upon the contents of the AC. Finally, it is able to direct which of its own instructions are to be carried out next.

Both the AC and each memory word are assumed capable of holding a single integer, no matter how large. Although the RAM has only one register, it has an infinite number of words in its memory. In the figure on the next page, there is a communication line between the RAM's control unit and each of its memory words. This symbolizes the fact that as soon as an address is specified by the RAM program, the corresponding word of memory is instantly accessible by the control unit.

Although a great variety of RAMs have been defined by various authors, we are content with a fairly simple sort which is programmable in the language shown in the table below. We shall call it the *random access language* (RAL):

Mnemonic	Argument	Meaning
LDA	X	Load the AC with the contents of memory address X.
LDI	X	Load the AC indirectly with the contents of address X.
STA	X	Store the contents of the AC at memory address X.
STI	X	Store the contents of the AC indirectly at address X.
ADD	X	Add the contents of address X to the contents of the AC.
SUB	X	Subtract the contents of address X from the AC.
JMP	X	Jump to the instruction labeled X.
JMZ	X	Jump to the instruction labeled X if the AC contains 0.
HLT		Halt.

When we speak of the contents of memory address X, we refer, of course, to the integer currently stored in the memory word whose address is X. To load the AC *indirectly* with the contents of memory address X means not to load the contents of address X, but to treat those contents as yet another address (whose

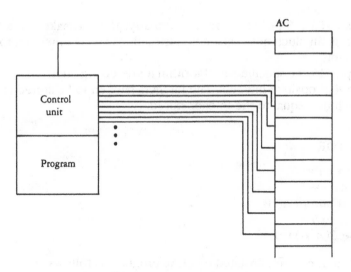

Figure 17.1   A random access machine

contents are to be loaded). The instruction STI uses the same kind of indirection, but in reverse.

When a RAL program is written out (see the example below), we may number its statements so that the jump commands (JMP and JMZ) are clearly understood. Thus, JMP 5 means that the next instruction to be executed is instruction 5. But JMZ 5 means to execute instruction 5 next if the AC contains 0, otherwise, to execute the next instruction in the program.

Indirection is a very useful feature of RAM programs. Both LDI and STI use indirection in the following manner: "LDI X" means first to look up the contents of X. If the integer stored there is Y, then the RAM is next to look up the contents of Y and, finally, to load that integer in the AC. This sort of indirect memory reference also occurs in the STI instruction type: Look up the contents of X, get the integer stored there, say Y, and then store the contents of the AC at Y.

As in the case of Turing machines, no attention is paid to input or output in RAMs; since they are not "real" machines, there is no point in trying to communicate with a user. Instead, a certain finite initial set of integers is assumed to exist already in certain memory words, with all the rest containing 0s. At the end of a RAM's computation what remains in memory is considered to be its output.

A RAM halts under one of two conditions. If execution comes to a HLT command, all operations cease and the current computation comes to an end. If the RAM comes to a nonexecutable instruction, execution also ceases. A nonexecutable instruction is one whose arguments make no sense in the context of the RAM pictured above. For example, no memory location has a negative

111

address, so STA −8 makes no sense. Similarly, JMP 24 makes no sense in a 20-line program. Such nonexecutable instructions are assumed not to exist in RAL programs.

In Chapter 15, we specified an algorithm for detecting duplicates in an input sequence A of positive integers. The integers happen to be stored in array B, each at an index equal to the integer stored.

STOR

for $i \leftarrow 1$ **to** n **do**
    if $B(A(i)) \neq 0$
        then *output* $A(i)$;
            exit
        else $B(A(i)) \leftarrow 1$

This algorithm can be translated to a RAL program as follows:

STOR · RAL

```
 1. LDI  3   /get ith entry from A
 2. ADD  4   /add offset to compute index j
 3. STA  5   /store index j
 4. LDI  5   /get jth entry from B
 5. JMZ  9   /if entry 0, go to 9.
 6. LDA  3   /if entry 1, get index i
 7. STA  2   /and store it at 2.
 8. HLT      /stop execution
 9. LDA  1   /get constant 1
10. STI  5   /and store it in B
11. LDA  3   /get index i
12. SUB  4   /subtract limit
13. JMZ  8   /if i = limit, stop
14. LDA  3   /get index i (again)
15. ADD  1   /increment i
16. STA  3   /store new value of i
17. JMP  1   /go to first instruction
```

The STOR program is more readily understood by referring to Figure 17.2 in which the program's use of both the AC and memory is clearly laid out.

The first five words of memory are devoted to variables and constants used by the program. In this particular case, the next four words contain a set of four integers to be tested for duplicates. These words are referred to collectively as

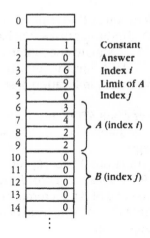

Figure 17.2 Memory layout for the STOR program

A, effectively an array. All the remaining words of memory, from 10 onward, comprise an infinite array *B* in which STOR will place a 1 each time it processes a new integer in *A*.

The first memory location contains the constant 1 which is used by STOR to increment the index *i*. This index, or rather its current value, is stored in location 3. A second index, *j*, is stored in location 5. It serves to access locations in *B*.

The first three instructions of STOR are

1. LDI 3
2. ADD 4
3. STA 5

These instructions look up the contents of the location we call *i*, add the contents of location 4, and then store the result in location 5. Specifically, the LDI instruction looks up the contents of location 3. Initially, the number stored here is 6, the first location in the array *A*. The number stored in location 6 is a 3 and *this* number is placed in the AC by the LDI instruction. Location 4 contains the constant 9 which delimits the *A* section of memory. The effect of adding 9 to 3 in the AC is to calculate the third location of *B*, namely 12. The computer then stores this number in location 5. Thus STOR retrieved 3 as the first integer of *A* and computed the location of the third member of *B*.

The next instruction of STOR, namely LDI 5, loads the number stored in the third location of *B* into the AC. This number happens to be 0 so at instruction 5, execution jumps down to instruction 9. Instructions 9 and 10 retrieve the con-

113

stant 1 and store it in the third location of B: in reality, this happens to be 12 and that is where STOR places a 1. This indicates that STOR has just encountered a 3. If (given another A sequence) STOR should encounter a 3 later, it would have a 1 in the AC when it reached the JMZ 9 instruction. In such a case, it would immediately load the current value of i in location 3, store it in location 2 and then halt.

Location 2 contains the answer when STOR terminates. If 0, it means that no duplicates were found among the integers of A. Otherwise it will contain the location in A of a repeated integer.

The rest of the program, instructions 11 to 16, retrieves the index i from location 3 and then subtracts 9 (in location 4) to decide whether STOR has reached the end of A. If not at the end of A it increments the index i and stores it again in location 3 before jumping back to the head of the program in order to test the next integer in the A sequence.

At first glance, it would appear that RAMs are more powerful than Turing machines in terms of the functions which they compute, but this is not really true. In Chapter 66, it will be shown that Turing machines can do anything that RAMs can do. In the meantime, it would be useful to check the converse. It may seem obvious, but can RAMs *really* do anything that Turing machines can do? Luckily, it is not hard to prove. Given an arbitrary Turing machine with a semi-infinite tape (see Chapter 31), use the RAM memory to simulate the tape as shown in Figure 17.3.

Given this one-to-one correspondence between words of RAM memory and squares of the Turing machines tape, it is now only a matter of writing a RAM program to mimic the Turing machine's program. Each quintuple of the latter is

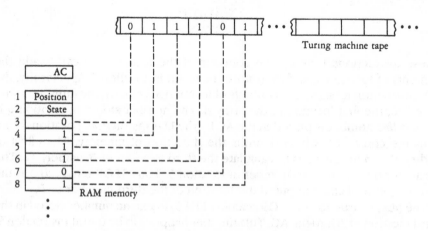

Figure 17.3 A RAM simulating a Turing machine

replaced by a number of RAL instructions which have the same effect on the RAMs memory as the quintuple would on the Turing machine tape. To this end, it is necessary for the RAM program to remember the position of the Turing machine's read/write head as well as its current state.

Because of their equivalence with Turing machines and, indeed, because of their equivalence with all the most powerful known abstract computational schemes, RAMs offer an alternate vehicle in the study of feasible computations. Some questions are more easily asked and answered in the framework of such a model of computation; of all the schemes, it is most like a standard, digital computer. It is also one of the most convenient schemes for expressing actual computations such as the RAL program we have just examined.

Problems

1. Alter the RAL program STOR so that when a computation is finished and the input sequence contained a duplicate integer, we know what integer that was.

2. Complete the proof that a RAM can carry out any Turing machine computation.

3. Our RAM language RAL contains no instruction for multiplication or division. Write a RAL program called MPY which takes two integers X and Y as input in addresses 1 and 2, respectively, and outputs their product in address 4.

References

D. E. Knuth. *The Art of Computer Programming,* vol. 1: *Fundamental Algorithms.* Addison-Wesley, Reading, Mass, 1969.

A. V. Aho, J. E. Hopcroft, and J. D. Ullman. *The Design and Analysis of Computer Algorithms.* Addison-Wesley, Reading, Mass. 1974.

SPLINE CURVES

Smooth Interpolation

To watch a plotter driven by a spline curve program (Figure 18.1) is to enter a magic realm where simple algebra springs to life. Several prominent points on the plotter paper are laced through, one after another, by a sinuous, natural-looking curve. The plotter pen never hesitates. As it approaches one point, it moves to the left, looking for a moment as though it will miss the point entirely. But then it heads more and more directly toward the point, finally entering from this unexpected direction. Suddenly, the reason for the circuitous approach is clear—the next point lies well to the right of the point just crossed.

In computer graphics, data analysis, and many other applications, one wants a "natural" curve to connect a number of points. The graphic application, for example, could well involve the drawing of a figure defined by a collection of points. The profile of a face might be initially generated by a computer as a set of 25 points. A spline-drawing program then connects the dots, so to speak, by a curve that is nothing if not facelike. In many situations, scientists measure a continuously varying quantity such as air pressure, response time, or magnetic field strength by a relatively small set of discrete measurements. The value of

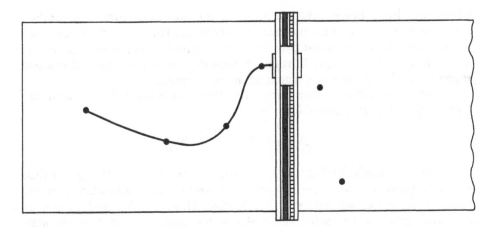

Figure 18.1 A plotter draws a spline curve

the quantity between measurements is estimated by interpolation; a spline curve that (hopefully) matches the unmade measurements is drawn. Thus the data are fleshed out to an entire curve.

The points to be connected are called *control vertices.* They can be literally anywhere in the plane, and naturally they all have coordinates:

$$(x_0, y_0), (x_1, y_1), \cdots, (x_{n-1}, y_{n-1}), (x_n, y_n)$$

A spline program connects the points in consecutive pairs by a function that has the form

$$f(t) = (x(t), y(t))$$

In other words, the spline function f is really two separate functions $x(t)$ and $y(t)$. The independent variable t is called a *parameter*; as t ranges from an initial value of t_0 to a final value of t_1, the curve is drawn as a much finer succession of points $(x(t), y(t))$. Thus the points on the curve arise by computing both functions for each t value and then plotting the resultant point. Since functions x and y need have no relationship whatever to each other, they can be studied separately. In what follows, for example, we examine $x(t)$ alone, knowing that the theory and techniques developed apply equally well to $y(t)$.

Among all the candidate functions for $x(t)$, the polynomials are the simplest and easiest to compute. They also offer some promise of flexibility, sinuosity, and that elusive property "naturalness." But what polynomial should we use?

117

Obviously a linear polynomial is too simple; it produces angular, connect-the-dots style curves. What about quadratic polynomials? They are certainly capable of curvature. It will be clear presently why second-degree polynomials are simply not flexible enough. It has to do with the manner in which individual segments of the spline curve match up at the control points.

The next possible candidate for the function $x(t)$ is the cubic, or third-degree, polynomial. Its general form is

$$x(t) = a + bt + ct^2 + dt^3$$

where t runs, let us say, from 0 to 1. To be more specific, we will suppose that x is in the process of determining first coordinates of curve points between the control vertices (x_i, y_i) and (x_{i+1}, y_{i+1}). What values of a, b, c, and d will take the curve from the ith to the $(i+1)$th control point? To find out, one may substitute $t = 0$ and $t = 1$ into the formula above. Since $x(0) = x_i$ and $x(1) = x_{i+1}$, two conditions immediately arise:

$$x_i = a \tag{1}$$

$$x_{i+1} = a + b + c + d \tag{2}$$

Obviously, there is considerable choice left to us for the coefficients b, c, and d. Any combination of b, c, and d yielding the sum $x_{i+1} - a$ would appear to be suitable. Yet the segments of the curve must do more than merely meet each other at the control vertices; they must do so smoothly. The curve should not appear to bend at these vertices. It makes sense, then, to require that the spline segments have the same slope at (x_i, y_i). The most direct approach to this requirement is to specify slope as giving rise to two more conditions. Since the derivatives of $x(t)$ at $t = 0$ and $t = 1$ must be s_i and s_{i+1}, respectively, we merely substitute $t = 0$ and $t = 1$ into the derivative of the spline function to obtain

$$s_i = b \tag{3}$$

$$s_{i+1} = b + 2c + 3d \tag{4}$$

Armed with conditions (1) to (4), we can now bend the spline function to our will. These conditions amount to four linear equations in four unknowns. Their solution is straightforward:

$$a = x_i$$
$$b = s_i$$
$$c = 3(x_{i+1} - x_i) - 2s_i - s_{i+1}$$
$$d = 2(x_i - x_{i+1}) + s_i + s_{i+1}$$

Given the quantities x_i, x_{i+1}, s_i, and s_{i+1}, it is now possible to find the coefficients for the cubic curve comprising the ith segment of the spline. These coefficients may as well be called a_i, b_i, c_i, and d_i.

One could almost leave the subject of splines at this point except that a few decisions have yet to be made by a user of the technique: How are the slopes s_i chosen? At what values of t does the curve pass through the control vertices?

First, the derivatives s_i may be chosen by a variety of means: They can be guessed by visual inspection of the control vertices. A more objective technique is to fit a simple quadratic curve to the three coordinates x_{i-1}, x_i, and x_{i+1}. This curve is not in the plane of the spline curve but in a different plane that could be called the *parameter space*. The three points (t_{i-1}, x_{i-1}), (t_i, x_i), and (t_{i+1}, x_{i+1}) determine a parabola. It can easily be found by the process of substitution and the solution described above; merely start with a general quadratic formula like $a + bt + ct^2$.

But what values of t are supposed to give rise to the coordinates x_i? Here again a number of approaches suggest themselves. In the technique called *uniform cubic splines*, the control values of t_i are uniformly spaced along an interval. Suppose, for example, that t runs from 0 to 1. If the spline curve is to have n segments, then the ith segment would be drawn as t varied from $(i - 1)/n$ to i/n. Of course, the values of a_i, b_i, c_i, and d_i are strongly affected by the domain that t varies within in generating the ith segment. Thus the coefficients must be recomputed with the values $(i - 1)/n$ to i/n in place of the 0 and 1 used earlier. But the procedure is exactly the same. The uniform approach breaks down, in a sense, when there are large gaps between certain consecutive pairs of control vertices. In such a case, one may choose a nonuniform spacing based on the euclidean distance between successive vertices.

The surface of splines has barely been scratched here. For example, our analysis of $x(t)$ can be extended not only to $y(t)$ but also to a third coordinate, $z(t)$. In this manner a spline curve in three-dimensional space is generated. Other techniques arise from requiring that both first and second derivatives match at the control vertices.

Problems

1. Given three vertices (t_1, x_1), $(t_2\ x_2)$, and (t_3, x_3), find the equation of a parabola that passes through them.

2. Is interpolation by quadratic splines possible? How does the situation change when one tries to develop the equations involving x_i, x_{i+1}, s_i, and s_{i+1}?

3. Write a uniform cubic-spline program that takes n control vertices as input and produces the coefficients for each segment as output. On a screen or plotter, use the program to plot first the control vertices and then the curve. The curve itself will be a succession of minute line segments.

References

William H. Press, Brian P. Flannery, Saul A. Teukolsky, and William T. Vetterling. *Numerical Recipes: The Art of Scientific Computing.* Cambridge University Press, Cambridge, 1986.

Bruce A. Artwick. *Microcomputer Displays, Graphics, and Animation.* Prentice-Hall, Englewood Cliffs, N.J., 1985.

19

COMPUTER VISION

Polyhedral Scenes

The way in which we humans perceive, analyze, and classify the images we confront in our everyday world continues to be a mystery. Although the retina of the eye and the first tier of cells in the visual cortex appear to decompose a viewed image into line segments, edges, and other simple pictorial elements, we know almost nothing about how the brain analyzes the scenes which it confronts, how it spots an entranceway, or how it recognizes the face of a friend.

How, for example, does the human brain make sense of the image in Figure 19.1? Before we describe it to ourselves as "an archway on a platform," "part of an ancient temple," or anything else, we assuredly have recognized that the structure represented by this picture is composed of five separate elements: two horizontal slabs, two vertical bars, and a horizontal one. At least this is a reasonable interpretation, and there is reason to suppose that such a preliminary analysis forms an essential part of our internal recognition process.

The image shown here represents an example of what could be called a *polyhedral scene,* that is, an assembly of solids each of which is bounded by plane faces. The faces of these solids meet along straight-line segments having a

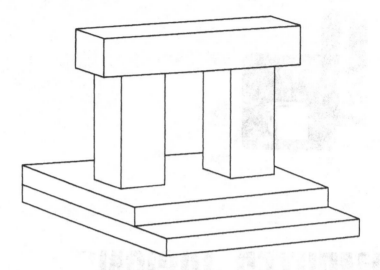

Figure 19.1 Part of an ancient temple?

characteristic geometry and showing only a finite number of relationships where two or more of them meet. For example, in the picture above, there are essentially only five types of such relationships. These are shown in Figure 19.2 in schematic form and provided with descriptive names which we shall use later.

It is fascinating to think that it is possible to analyze images of polyhedral scenes by computer and, based only on information about junctions of the sort shown here, come to an automatic decision about the separate solids which compose the scene. A major step in such an analysis was taken in 1971 by David Huffman and Maxwell Clowes. Their work was soon extended by David Waltz at the Massachusetts Institute of Technology. Initially, we describe a cut-down version of Waltz's approach in which scenes without shadows are analyzed.

To successfully use the junction types of Figure 19.2 in a scene analysis program, they must be labeled according to the kind of edges or cracks that give

Figure 19.2 Lines meet in only five ways.

rise to the line segments which compose them. In the kinds of scene (temporarily) under consideration, only four kinds of label may be attached to line segments. Each of these corresponds to a specific physical reality.

$-$	$+$	$\longrightarrow$	C
Concave edge	Convex edge	Obscuring edge	Crack

For example, a concave edge is a straight line along which two faces meet at an angle of less than 180° relative to the viewer.

By using such labels, it is now possible to develop an extended list of junctions in terms of the sort of edges or cracks which meet at them as in Figure 19.3. This extended list is not tremendously longer than the original one because the physics and geometry of possible structures place very severe constraints on what edges or cracks may meet at a specific type of junction. For example, it is difficult to see how a fork junction could arise as the meeting of three obscuring edges, that is, edges each of which represents the visual boundary of a particular solid within the scene.

This list of junction labels is somewhat incomplete and specifically excludes certain degenerate or unlikely configurations. However, even with such a list the success of Waltz's analysis program is virtually ensured. The reason for this lies in the interaction of junctions which share a common line segment: it does not change its physical identity in going from one junction to the other.

For example, if a fork shares a line segment with a T, forming its "upright," there are many a priori possible combinations. Of these, only 4 may actually occur as far as the list of Figure 19.3 is concerned:

$$(F2, T1), (F2, T5), (F3, T3), (F3, T4)$$

Waltz's program takes advantage of this sort of interaction. Starting with a list of all possible labels at each junction, the program selects a specific one, considers a neighboring junction, and reduces the list of possibilities for both junctions by throwing away those labels at one junction which cannot be matched at the other. In traveling to a third junction, this effect is often enhanced by virtue of the second junction's list already being reduced by interactions with the first junction.

As an illustration of this general scheme, we reexamine the image of our original polyhedral scene and place markers A, F, L, T, or K beside each of its junctions. We begin the analysis at an arbitrary junction—the one marked with an arrow in Figure 19.4.

This junction is a fork and initially has three labels, $F1$, $F2$, $F3$, on its list. Below the fork is an arrow junction with the labels $A1$ and $A2$ on its list. When an

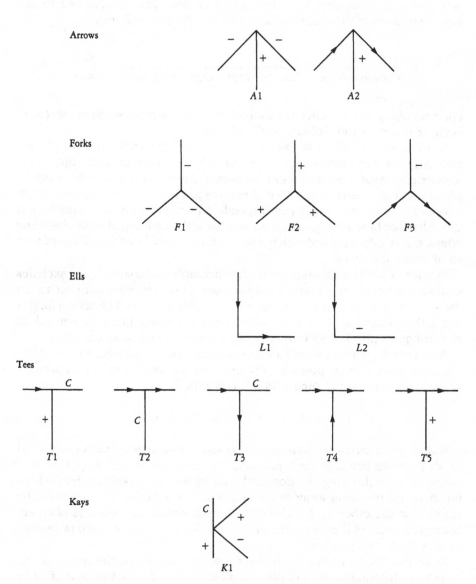

Figure 19.3 There are just 13 labeled junctions.

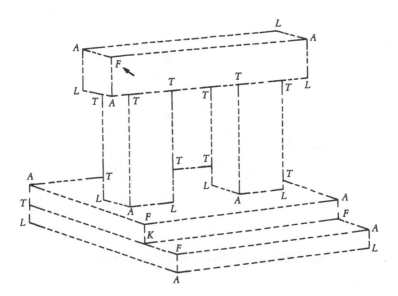

Figure 19.4 The analysis begins.

attempt is made to label the line segment connecting these two junctions, however, the program discovers that only + will work, resulting in a reduction of the fork junction's list from three possibilities to just one, namely F2. So far, the arrow junction's list has not been reduced at all: The two edges forming the head of the arrow could be both obscuring or both concave edges as far as the program knows. However, in next considering the adjacent T junctions, the program quickly recognizes that there is no way to match A1, the arrow with concave edges, with any of T1 through T5: The latter list does not allow concave edges. This forces a reduction in the arrow list to just one member, A2, and the current situation can now be portrayed as shown in Figure 19.5.

When the program comes to examine the T junction to the left of the A2 junction, it is unable to reduce the list (T1, · · ·, T5) at all since each T on this list has an arrow label on one of its bars. If, however, at some later point the program analyzes the L junction adjacent to this T, it attempts to reduce its list (L1, L2) but is unsuccessful by using information about adjacent junction labels alone. This particular ambiguity is removed by a special device of Waltz's, that is, to recognize that this particular junction lies on the visual boundary of the entire image. It therefore cannot be incident with any concave edges which also form part of that boundary. This eliminates L2 from the list.

Proceeding in this way from junction to junction, the program eventually arrives at the set of edge labels for the entire image shown in Figure 19.6.

125

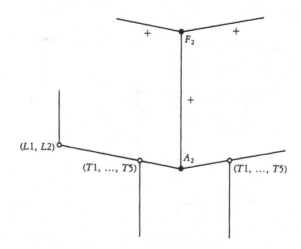

Figure 19.5 The analysis determines the first two junctions

The Waltz program attempts to output a labeling of all the line segments in the image of a given polyhedral scene. Our restricted version of his program would output the sets shown here. In all but 15 cases, the line segments have been given unique labels. In the remaining cases, our program cannot decide between certain alternatives. Indeed, some of these ambiguities are perfectly

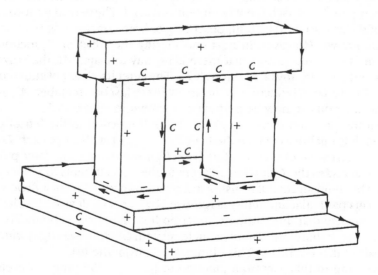

Figure 19.6 The analysis complete except for ambiguities

reasonable given our restricted program's "knowledge" of polyhedral scenes in general or this scene in particular. For one thing, the program knows nothing of support relationships, and rather than decide that the bottoms of the pillars are concave edges, it presumes that they could equally well be obscuring edges; why should the pillars not be floating a few inches above the platform? For another thing, even we do not know whether the outermost ambiguous edges of the lintel should be cracks or obscuring edges!

Waltz's complete program takes shadows into account and operates on a much larger set of junctions (and their labels) which results from the inclusion of shadow edges, symbolized by an arrow crossing the appropriate line segment toward the shadow:

$$\begin{array}{r} \text{Light} \\ \hline \text{Shade} \end{array}$$

The earlier workers in polyhedral scene analysis had supposed that shadows in such scenes were merely a confusing irrelevancy. It was Waltz who demonstrated their usefulness in removing ambiguities of the sort which we have already encountered in our example. In Figure 19.7 there are four shadow regions, and these give rise to three new junction labels (Figure 19.8). Of course, once shadows are taken into consideration, many more junction labels than merely these three are possible. However, even with a somewhat ex-

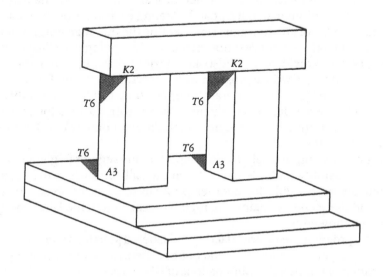

Figure 19.7 Shadows remove ambiguities

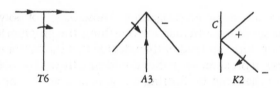

Figure 19.8 Shadow labelings

panded list, the presence of the shadows in our polyhedral scene is enough to resolve all but two of the previous ambiguities over edge labels! Note that new junction labels are introduced only where a shadow line (as opposed to a shaded obscuring edge) meets a junction or creates a new one.

The underlying algorithm in Waltz's program is best understood in the context of exactly what goes into it and what comes out. The input to Waltz's program is not the digitized image of some polyhedral scene, but rather a list of its junctions, line segments, and regions. Although such a list is not very difficult to generate given

A crisp, well-lighted polyhedral scene
A digitizing camera, interface, and computer
An effective line-finding program

Waltz assumed, for each of the polyhedral scenes he studied, the existence of such a facility and simulated it by hand. Naming each junction, he created a list of them and a list of junction pairs representing the line segments joining them. It is interesting that such a description is entirely topological, since no distance information is provided. It is also not terribly difficult, under the first two conditions above, to write a program which determines the uniform regions (in the digitized image of a polyhedral scene) and their brightness. For this reason, Waltz felt that it was fair to supply his program with information about which line segments bounded each region, including the outer one(s) surrounding the objects represented.

The program outputs a label for the inputted line segments, giving each one a single or possibly a multiple label. When such a labeling is available, it would presumably be possible for another program to identify objects or parts of objects in the scene by coalescing regions according to the labels assigned to their bounding edges.

Waltz's program uses a large data base of labeled junctions and line segments, as well as various selection rules and optional heuristics, to aid in the systematic elimination of impossible combinations of labels assigned to junctions and line segments. As such, it is quite a large program and is impossible to describe here in real detail. Nevertheless, the most interesting and important part of the program, the one which eliminates labels for junctions, can be summarized as follows:

At each junction the program carries out three steps:

1. It creates a list of all possible labels for a junction of that type.

2. It examines adjacent junctions, using their current label lists to restrict the possibilities at this junction. Impossible labels are removed from its list.

3. Using the reduced list of labels thus obtained, it removes any newly impossible labels from adjacent junctions and continues to propagate such restrictions outward as far as they occur.

Waltz showed that the solution found by his program is always the same no matter at which junction it begins. The program is also guaranteed to terminate after executing the steps above at each junction in the image.

After demonstrating the effectiveness of his program on relatively simple polyhedral scenes, Waltz was able to extend it to handle certain degeneracies and accidental alignments, increasing its power still further. But how powerful is such a program ultimately, and what does it tell us about the possibilities for "seeing computers"? First, ordinary scenes which we humans deal with are far from being polyhedral (with the possible exception of certain urban land-scapes). Does Waltz's program appear to be extendable to such a complicated visual realm? Probably not. However, manufacturing settings already do have, or can be made to have, enough geometric simplicity to allow such a program to operate effectively as part of the visual software of a manipulating robot.

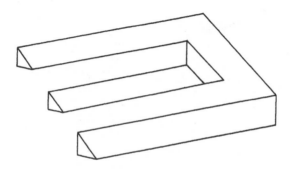

Figure 19.9 A problem for the program

Problems

1. Show that with the junction labels supplied here Waltz's program would be unable to decide upon any labeling for Figure 19.9.

2. Considering all "possible" polyhedral scenes, those involving constructable polyhedral solids enjoying the usual gravitational support relationships, how many additional possible labels can you find for the arrow junction?

3. Assuming the creation of eight shadow junctions in the last archway figure, demonstrate that all but two of the remaining ambiguities of edge labels are cleared up. What will be the final edge labels outputted by the program in this case? Is the program's failure to resolve the two ambiguities understandable?

References

Patrick Henry Winston. *The Psychology of Computer Vision.* McGraw-Hill, New York, 1975.

Patrick Henry Winston. *Artificial Intelligence.* (3rd edition) Addison-Wesley, Reading, Mass., 1992.

20

KARNAUGH MAPS

Circuit Minimization

Besides computers, our technological society is replete with devices that require logical control: vending machines, automobile ignition and carburation systems, automatic tellers, and elevators, to name just a few. Consider the case of elevators. An elevator control device receives floor requests as inputs and generates commands to the hoist and door motors as outputs (Figure 20.1). How complicated is the essential control function? It turns out to be relatively simple.

The problem of minimizing logical circuitry is called *boolean minimization,* and one of the best techniques for solving small instances of this problem is the *Karnaugh map.* Such maps provide easily inspected, two-dimensional visualizations of boolean functions and lead directly to a short formula for the function; the shorter the formula, the simpler the circuit (Chapter 13). Most useful for functions of two, three, and four logic variables, Karnaugh maps become steadily more unwieldy beyond these numbers.

Specifically, a Karnaugh map is an array of cells, and each cell stands for a

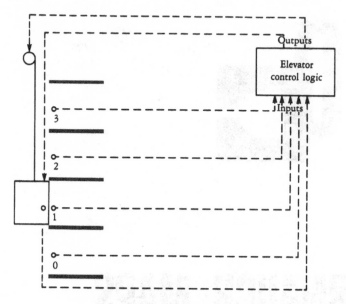

Figure 20.1 An elevator system

specific product of variables and their complements. For example, the two-variable map has four cells standing for $x_1 x_2$, $x_1' x_2$, $x_1' x_2'$, and $x_1 x_2'$.

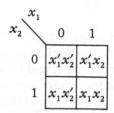

Each row and column of the map correspond to a value of one of the two logic variables. These values are assigned in a way that produces a 1 when they are substituted into the product lying at the intersection of their respective row and column. Thus the $x_2 = 1$ row intersects the $x_1 = 0$ column at $x_1 x_2'$, and at these values $x_1 x_2' = 1$. The reasons for this arrangement will become apparent when we show how two-variable maps are used.

Consider the function $f(x_1, x_2)$ written in disjunctive normal form as $x_1 x_2' + x_1' x_2 + x_1 x_2$. For each product appearing in this formula we place a 1 in the corresponding cell of the two-variable Karnaugh map.

$$f(x_1, x_2) = x_1 x_2' + x_1' x_2 + x_1 x_2$$

	x_1	
x_2	0	1
0	0	1
1	1	1

Notice that adjacent cells correspond to products with common factors. For example, the two bottom cells correspond to $x_1 x_2'$ and $x_1 x_2$. These products have the common factor x_1, and we may write

$$x_1 x_2 + x_1 x_2' = x_1(x_2 + x_2')$$
$$= x_1$$

obtaining a simplification in part of the formula for f. Accordingly, we link the 1s in these two cells by a rectangle. We also link the two vertical 1s in the same manner.

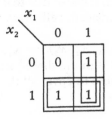

Now

$$
\begin{aligned}
f(x_1, x_2) &= x_1 x_2' + x_1' x_2 + x_1 x_2 \\
&= x_1 x_2' + x_1 x_2 + x_1' x_2 + x_1 x_2 \qquad \text{(repeating a product} \\
& \qquad\qquad\qquad\qquad\qquad\qquad\quad \text{changes nothing)} \\
&= x_1(x_2' + x_2) + (x_1' + x_1)x_2
\end{aligned}
$$

To reach the same conclusion but avoid all this algebra, the rule used in Karnaugh maps is simply to replace each rectangle by the constant factor in the corresponding products: The bottom rectangle stands for $x_1 x_2'$ and $x_1 x_2$, the constant part of which is x_1. The constant part of the vertical rectangle is x_2.

Three-variable Karnaugh maps are twice as complicated as two-variable maps. Here the cells correspond to all possible products of the three variables x_1, x_2, and x_3 and their complements.

133

x_1 \ x_2x_3	00	01	11	10
0	$x_1'x_2'x_3'$	$x_1'x_2'x_3$	$x_1'x_2x_3$	$x_1'x_2x_3'$
1	$x_1x_2'x_3'$	$x_1x_2'x_3$	$x_1x_2x_3$	$x_1x_2x_3'$

In this map, rows correspond to values of a single logic variable x_1, but columns correspond to pairs of values for x_2 and x_3. Again, these values are assigned in such a way that the product at the intersection of a given row and column takes on the value 1 when the corresponding substitutions are made into the product. At the same time, the paired values assigned to columns have the interesting property that only one bit changes when we go from one pair to the next. This is true even when we go from the last pair back to the first. Such a sequence is a very simple example of a Gray code. If one examines any pair of adjacent cells in this map, one sees that the corresponding products have two variables in their common part. It is also true that the products residing within a square configuration of four adjacent cells have one variable in their common part.

Consider the function $f(x_1, x_2, x_3) = x_1'x_2x_3 + x_1'x_2x_3' + x_1x_2x_3 + x_1x_2x_3' + x_1x_2'x_3'$. When we fill in the three-variable map above with 1s from this function, we find it possible to cover all the 1s with two rectangles.

x_1 \ x_2x_3	00	01	11	10
0	0	0	1	1
1	1	0	1	1

One of the rectangles "wraps around," embracing both the right- and left-hand 1s on the bottom row. As in the two-variable map, we extract the common part of all products contained in each rectangle. The square results in the expression x_2, and the horizontal rectangle results in x_1x_3'. It follows that

$$f(x_1, x_2, x_3) = x_1x_3' + x_2$$

The reduction resulting from the square configuration of cells can easily be confirmed algebraically as follows:

$$x_1'x_2x_3 + x_1'x_2x_3' + x_1x_2x_3 + x_1x_2x_3' = x_1'x_2(x_3 + x_3') + x_1x_2(x_3 + x_3')$$
$$= x_1'x_2 + x_1x_2$$
$$= (x_1' + x_1)x_2$$
$$= x_2$$

Four-variable Karnaugh maps are twice as complicated as three-variable maps. The products are so large that it is more convenient to number them according to the combination of values by which they are indexed. For example, instead of writing $x_1 x_2' x_3' x_4$, we write 1001. In fact, we go one better than this and write the decimal equivalent, namely, 9.

x_1x_2 \ x_3x_4	00	01	11	10
00	0	1	3	2
01	4	5	7	6
11	12	13	15	14
10	8	9	11	10

By using the same Gray code sequence on both rows and columns, the desirable reduction properties of adjacent cells are preserved: One can have 2, 4, 8, or even 16 adjacent cells in various rectangular configurations. An attempt is made to find the smallest number of rectangles which cover all the 1s for a given function. Then the corresponding reduced expressions are written down in sum form; this is the reduced form of the function.

Returning now to the elevator example, we analyze the logic operations required and use the resulting four-variable function to fill in a Karnaugh map.

Encode request for service from each of four floors by a 2-bit binary number as follows:

X_1	X_2	
0	0	Ground
1	0	Floor 1
0	1	Floor 2
1	1	Floor 3

Encode current elevator position in the same way;

X_3	X_4	
0	0	Ground
1	0	Floor 1
0	1	Floor 2
1	1	Floor 3

Output of the control circuit includes commands for the elevator to go up *(u)*, down *(d)*, and to stop *(s)*. In this example, we develop the "up circuit" only. The four-variable Karnaugh map is filled in directly by inspection. Each cell is examined, and if the current position is below the floor requested, a 1 is placed in that cell; otherwise, a 0.

No more than three rectangles suffice to cover all the 1s in this map, and the resulting expression for *u* is

$$u(x_1, x_2, x_3, x_4) = x_1 x_3' + x_2 x_3' x_4' + x_1 x_2 x_4'$$

Although the resulting expression is in minimum disjunctive form, further algebraic simplification is possible:

$$x_1 x_3' + x_2 x_3' x_4' + x_1 x_2 x_4' = x_1 x_3' + x_2 x_4'(x_3' + x_1)$$

The corresponding circuit has just seven logic gates (Figure 20.2):

As we have already seen, the size of a Karnaugh map doubles with each added

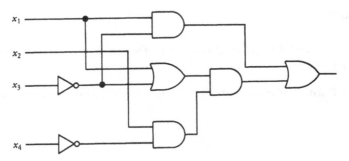

Figure 20.2 The up-control circuit

variable. Because of this and because of the increasing complexity of available configurations, the method is not used much when more than six variables are involved. It should be stressed, moreover, that Karnaugh maps are intended for the exclusive use of human designers. Because of our ability to take in large amounts of information visually, the method works more quickly than a process of scanning lines of algebraic formulas looking for common factors. At the same time, elaborate computer algorithms for the reduction of boolean functions have been developed.

Problems

1. Design a vending machine control circuit. A soft drink costs 45¢, and your circuit records the nickels, dimes, and quarters fed to the machine. It issues a command to dispense a can when the appropriate payment has been equaled or exceeded. Assume that at least one quarter is used and that coin counters make available binary counts for each kind of change.

2. Devise a five-variable Karnaugh map by putting two four-variable maps together in a certain manner. You may have to introduce a new kind of cell adjacency.

3. For each kind of map presented here, find a function which cannot be simplified at all. In each case, what proportion of all possible functions cannot be simplified?

Reference

M. M. Mano. *Digital Logic and Computer Design.* Prentice-Hall, Englewood Cliffs, N.J., 1979.

Charles H. Roth, Jr. *Fundamentals of Logic Design,* 2d ed. West, St. Paul, Minn., 1979.

21

THE NEWTON – RAPHSON METHOD

Finding Roots

From the earliest days, computers have been applied to the problem of solving one or more nonlinear equations. Such equations arise in a host of settings, from abstract inquiries into the general nature of solutions to complex physical systems described by one or more polynomial, trigonometric, or other well-behaved functions.

To solve an equation such as

$$x^3 \sin y + \cos (x^2 y + y) = 23xy \tan(xy)$$

one must, in effect, find the roots of an expression. After all, transposing one side of the equation to the other will produce zero on one side and an expression on the other. Any values of the independent variables that satisfy the original equation make the new expression equal to zero. Such values are called *roots* of the expression.

In the simplest, one-dimensional (single variable) setting, an equation like that in Figure 21.1 may become the focus of a root-finding investigation. This single-variable cubic function has just one root. The Newton–Raphson method finds the root by a simple, ingenious method that involves guessing the root, then improving the guess through a series of iterations.

In Figure 21.1, an initial guess, r_1, is obviously wide of the mark. The true root is found exactly where the curve crosses the x-axis. But the tangent to the curve at r_1 provides the clue to a better place to look. The tangent crosses the x-axis much nearer to the true root than r_1. The point of crossing, r_2, becomes the next guess in the iterative process. As most readers will have guessed by now, the whole process is repeated at the point r_2. The tangent to the curve at r_2 crosses the x-axis even closer to the true root than r_1 did.

When the Newton–Raphson method works, it works with incredible effi-ciency. The successive approximations $r_1, r_2, r_3, \ldots$ to the true root improve by a quadratic factor of precision. In other words, if e_i is the error in the ith

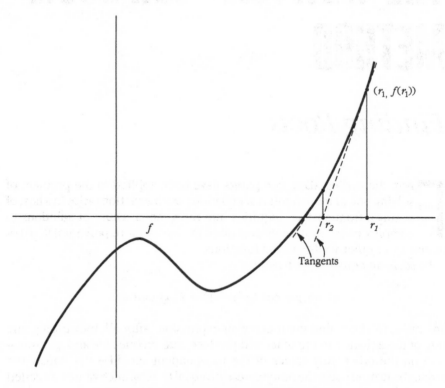

Figure 21.1 Newton–Raphson method converges

approximate root $r_i(e_i < 1)$, then

$$e_i + 1 = k \cdot e_i^2$$

where k is some constant. The number of significant digits in the approximation doubles with each iteration!

The basic algorithm is quite simple. It relies on the formula for the tangent to a smooth curve at a point, x. Students of calculus will understand that the derivative plays a key role in the formula. They will also understand that they must supply the derivative $f'(r)$ in the algorithm:

> **input** guess, limit
> $r \leftarrow$ guess
> **for** $j \leftarrow 1$ **to** limit
> $\quad r \leftarrow r - f(r)/f'(r)$

The *correction term*, $C = -f(r)/f'(r)$, when added to the old guess, happens to be the point at which the tangent to the curve at $(r, f(r))$ crosses the x-axis. Each time the value of this expression replaces r, the new value is much closer to the root.

In actual use, the Newton–Raphson method is often combined with other methods and certain special tests, which insure that convergence is proceeding properly. The method does not always work.

Figure 21.2 shows one way that the Newton–Raphson method can go astray. Suppose that one selected an initial guess, r_1, that was too far to the left of the true root. Here, the tangent to the curve at $(r_1, f(r_1))$ crosses the x-axis near the origin at the point r_2 which, though nearer the true root, is about to cause some misery.

On the very next iteration, the new approximate root, r_3, yields a new tangent and a new approximate root, r_4, which is not nearer to the true root but further away! In fact, depending on the actual value r_1 and the function f, the successive approximations may oscillate wildly back and forth forever, never approaching any fixed value.

The reason for this behavior lies in the difference between the method's local and global behavior. For any function that is well-behaved in the mathematical sense (i.e., not pathological), the choice of an initial guess, r_1, that is close enough to the root in question will guarantee convergence. To be fair, the method will quite often work even when the initial guess is some distance away from the true root. But in this context, "local" means "close enough to guarantee convergence." The user of the method, in other words, must have enough knowledge about the function f to make a good initial guess.

Systems of nonlinear equations may be solved by an extension of the

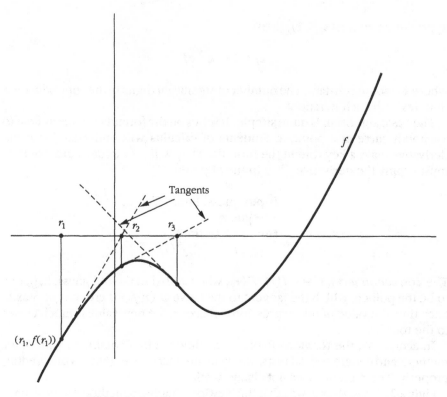

Figure 21.2 Newton–Raphson method fails to converge

Newton–Raphson method. In two dimensions, for example, successive guesses use the intersection points of directional tangent lines with the zero plane, as in Figure 21.3.

In cases of multivariable functions, the roots may have a more complicated structure. The "roots" of the function shown in Figure 21.3, for example, consists of a circle in the zero plane. Directional tangent lines correspond to partial directives in the x and y directions. The guessed root $r_1 = (x_1, y_1)$ will be improved upon as if x_1 and y_1 were the guessed roots in one-dimensional cross sections of the two-dimensional function, the standard formula being applied to each separately.

$$x_2 \leftarrow x_1 + f(x_1, y_1)/f_x(x_1, y_1)$$
$$y_2 \leftarrow y_1 + f(x_1, y_1)/f_y(x_1, y_1)$$

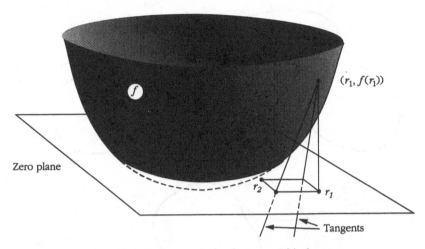

Figure 21.3 The root circle of a two-variable function

Here, f_x and f_y represent the partial derivatives in the x and y directions, respectively. The new guess, $r_2 = (x_2, y_2)$, should be closer to the true root than r_1 was.

The Newton–Raphson method may not work so easily with *systems* of nonlinear, multivariable functions.

$$f(x, y) = 0$$

$$g(x, y) = 0$$

The problem lies with the enormous complexities of the roots. Figure 21.4 illustrates what may go wrong.

Clearly, the user of this method must have special knowledge about the points where the zero curves of the two functions are likely to intersect. Once the neighborhood is known, however, the Newton–Raphson method can be extended. The correction terms C_j must now be found, however, by solving a simple matrix equation:

$$a_{11}C_1 + a_{12}C_2 = -f$$

$$a_{21}C_2 + a_{22}C_2 = -g$$

Here, the a_{ij} represent the partial derivatives. For example, a_{12} represents the partial derivative of f with respect to y. In this somewhat more compact notation, the four derivatives and the two functions, f and g, are all evaluated at the

143

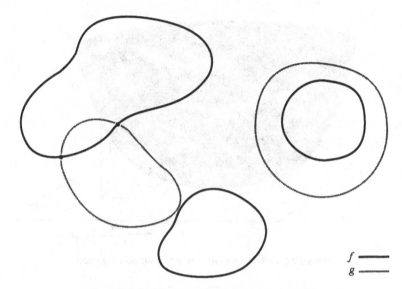

f ———
g -·-·-

Figure 21.4 The roots of f and g share two points

current, guessed root. When the matrix equation is solved for the correction factors, C_j, the new guess is readily found by adding the corrections.

In cases where the linear system has more than two variables and becomes unwieldy, the terms must be found by a solution algorithm such as the simplex method. (See Chapter 57.)

Problems

1. Define a function and a starting point for the Newton–Raphson method that will guarantee that the succession of approximate roots will diverge to infinity.

2. Attempt to compute $\sqrt{-1}$ by using Newton's method to locate the zero for the function $f(x) = x^2 + 1$. Make a graph of the function, pick an arbitrary initial value for x, and simulate the computation by hand. Why does it not converge?

References

Anthony Ralston and Philip Rabinowitz. *A First Course in Numerical Analysis,* 2nd ed. McGraw-Hill, New York, 1978.

William H. Press, Brian P. Flannery, Saul A. Teukolsky, and William T. Vetterling. *Numerical Recipes: The Art of Scientific Computing.* Cambridge University Press, New York, 1987.

22

MINIMUM SPANNING TREES

A Fast Algorithm

The subject known as graph theory is a branch of mathematics enjoying a special alliance with computer science in both its practical and theoretical aspects. First, the language, techniques, and theorems of graph theory may be applied to systems as diverse as data structures and parse trees. Second, graph theory itself is rich in problems which challenge our ability to solve by computer. Indeed, not many graph-theoretic problems appear to have algorithms that solve them in polynomial time. Many of the first problems shown to be NP-complete were problems in graph theory.

Among the well-solved problems is that of finding a minimum spanning tree for a graph. Specifically, given a graph G with edges of varying lengths, the problem is to find a tree T in G such that

1. *T spans G*, i.e. all G's vertices lie in T.
2. T has a minimum total length subject to condition 1.

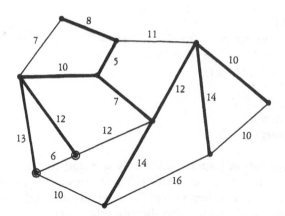

Figure 22.1 A graph and a spanning tree

The tree (shown in heavy lines) of Figure 22.1 spans the graph shown there, but it is not a minimum spanning tree. For example, if one of the edges incident with a circled vertex is removed from the tree and the edge joining the two circled vertices is added to it, then the resulting tree still spans the graph and has a shorter total length. Where is the minimum spanning tree? Is there more than one?

The most efficient known algorithm for the minimum spanning tree algorithm happens to be a greedy algorithm. Such algorithms solve optimization problems by optimizing at each step — like a greedy child confronted by a plate of cookies: each time the child is permitted to select a cookie, he or she always selects the largest (or tastiest) one.

The minimum spanning tree algorithm presented below was first developed in 1957 by R. C. Prim, an American mathematician. It proceeds by "growing" a spanning tree one edge at a time. Because the tree is to have a minimum overall length, the algorithm always selects the shortest available edge to add to the tree. In this sense the algorithm is greedy.

The algorithm, called MINSPAN, uses a list L of edges that join vertices currently in the tree under construction to those not yet spanned. The tree itself is denoted by T, and for each vertex v, E_v represents the set of all edges incident with v.

MINSPAN

1. select an arbitrary vertex u in the graph
2. $T \leftarrow \{u\}$
3. $L \leftarrow E_u$

4. $L \leftarrow sort(L)$
5. **while** T does not yet span G
   1. select the first edge $\{v, t\}$ in L
   2. $T \leftarrow T \cup \{v\} \cup \{v, t\}$
   3. $L' \leftarrow E_v - L$
   4. $L \leftarrow L - E_v \cap L$
   5. $L' \leftarrow sort(L')$
   6. $L \leftarrow merge(L, L')$

Initially, T consists of a single, arbitrarily selected vertex u. The list L at first consists of all the edges incident with u. These are sorted into order of increasing length. The algorithm then proceeds iteratively to add the shortest possible edges to the tree, one at a time. The shortest edge $\{v, t\}$ is easy to compute since it is always the first member of L; it joins a vertex t in T to a vertex v not in T. In steps 5.2, 5.3, and 5.4, the vertex v and edge $\{v, t\}$ are added to T, the list L' of new edges to be added to L is generated, and then L itself is reduced by all edges joining v to some other vertex in T. A new ordered list L is created at step 5.6 when the old ordered list L is merged with the sorted list L' of new edges.

It is interesting to examine the spanning tree generated by MINSPAN in specific cases. For example, MINSPAN produces the tree shown in Figure 22.2 in the case of the graph displayed in Figure 22.1. The initial vertex u is encircled.

In the analysis of algorithms, generally two steps are involved. First, the algorithm must be proved correct. Second, its complexity must be established as precisely as possible to within an order of magnitude. When an algorithm is being proved correct, the argument may be relatively informal. In proving programs correct, however, the precision of a specific language makes more

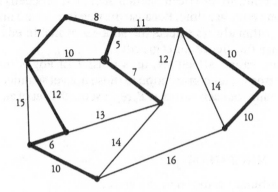

Figure 22.2 A minimum spanning tree found by MINSPAN

rigorous proofs possible (see Chapter 10). The same sort of remark can be made about establishing the time complexity of an algorithm (see Chapter 15). When an algorithm is given in more detail, it can be analyzed more deeply, sometimes resulting in a different order-of-magnitude complexity figure.

How do we know that MINSPAN really finds a minimum spanning tree for a graph G inputted to it? Suppose that MINSPAN outputs a spanning tree T in response to the graph G as input. If T is not a minimum spanning tree for G then G has some edge, $\{u,v\}$ say, that is shorter than any edge in T. At some point in its operation, MINSPAN included one of these two vertices in L while excluding the other. In either case, the edge $\{u,v\}$ was certainly shorter than any of the edges then joining L vertices to non-L vertices. Hence it must have been included at that time. This contradiction establishes the result.

Having produced a reasonably sound inductive argument that MINSPAN always finds a minimum spanning tree for a graph inputted to it, we now establish an order-of-magnitude figure to the time complexity of MINSPAN.

So far, we have not been very precise about what sort of data structures MINSPAN uses. It turns out to be most efficient to store both G and T as lists of edges according to the following format:

$$v_i: v_{i1}, c_{i1}; v_{i2}, c_{i2}; \cdots$$

In other words, using either an array or a linked list, we store a vertex v_i along with the vertices v_{ij}, where $\{v_i,v_{ij}\}$ is an edge of G and c_{ij} is the length of that edge. The analysis of time complexity now proceeds in stepwise fashion:

1. Selecting an arbitrary vertex of G will cost 1 unit of time.

2 & 3. Adding u to the (initially empty) list defining T costs 1 unit of time, and reading E_u into a list costs d_u, where d_u is the number of edges incident with u.

4. Sorting list L by some reasonably efficient method such as mergesort (see Chapter 40) costs $d_v \log d_v$ units of time.

5. Testing whether T spans G costs 1 unit of time.

   1. Since L is sorted into ascending order, the first edge of L is $\{v, t\}$, and this costs 1 unit of time to retrieve.

   2. The list defining T has no special order, and the new vertex and edge can be appended in 1 unit of time each.

   3. At this step the MINSPAN algorithm must examine each edge in E_v and decide whether it is in L. Since the edges of L are in sorted order, it costs $\log |L|$ units of time for each such decision. This yields a total of $d_v \log |L|$ units of time.

   4. Deleting the edges of $E_v \cap L$ from L, it is necessary once again to carry

149

out d_v searches of L for the members of E_v which happen to lie in L. Each search costs $\log |L|$ units, and each deletion costs 1 unit. Thus step 4.4 will cost $d_r \log |L| + 1$ units of time.

5. To sort L' costs no more than $d_v \log d_v$ units of time.

6. The amount of time taken to merge list L' with L would involve at most d_v searches and insertions in L, amounting to $d_v \log |L| + 1$ units.

This completes the detailed analysis of the algorithm. It is now necessary to "add all the figures" in a meaningful way. We label the vertices of G in the order that they appear in T, namely, v_1, v_2, v_3, and so on. Accordingly, their degrees are indicated by $d_1, d_2, d_3, \ldots$, and at the ith iteration of MINSPAN, L can be no larger than $d_1 + d_2 + \cdots + d_{i-1}$, while $|E_v| = d_i$. This leads to the following formulas:

Steps 1 to 4: $\qquad 2 + d_1 + d_1 \log d_1$

Steps 5 and 5.1 to 5.4: $\quad 4 + 2d_i \log (d_1 + \cdots + d_{i-1}) + d_i$

Steps 5.5 and 5.6: $\qquad d_i \log d_i + d_i \log (d_1 + \cdots + d_{i-1}) + d_i$

The final formula is obtained by adding the first formula to the iterated sum of the remaining two formulas:

$$2 + d_1(\log d_1 + 1) + \sum_{i=2}^{n} [4 + 3d_i \log (d_1 + \cdots + d_{i-1}) + 2d_i + d_i \log d_i]$$

With a small amount of simplification, this formula is easily shown to be bounded above by a much simpler expression, namely,

$$m \log 2m + 4n$$

where m is the number of edges in G and n is the number of vertices. Assuming that G is a connected graph, we have $m \geq n$, and it follows that the least upper bound is $O(m \log m)$. This gives the time complexity of MINSPAN as a function of the number of edges in G.

This completes the analysis of MINSPAN. Besides being a correct and quite efficient algorithm, it illustrates the basic simplicity and elegance of some of the best-solved problems in graph theory. The essential idea was simply to "grow" a spanning tree by adding one edge, the shortest available, at a time.

Interestingly enough, virtually the same algorithm can be used to find the maximum spanning tree. It is necessary only to alter instruction 4.1 to read, "Select the last edge $\{v, t\}$ in L." In the case of a closely related problem, that of finding the shortest path between two vertices, the situation is quite different. There is no way, apparently, to alter the shortest-path algorithm in order to find the longest path!

There is another minimum tree problem closely related to the one studied here. Suppose we are given a graph G and a specified subset S of G's vertices. What is the minimum-length subtree of G which spans all the vertices in S? Such a tree may certainly want to use some of the vertices of G not in S, but it is not, in general, required to span all the vertices of G. Such a minimum tree in G is called a *Steiner tree,* and the problem of finding a Steiner tree efficiently turns out to be much more difficult to solve than the minimum spanning tree problem. In fact, it appears to be intractible. (See Chapter 41.)

Problems

1. Select some vertex of G other than the one used as the starting point for MINSPAN in Figure 22.2. Is the minimum spanning tree that results different from the one in the figure?

2. Simplify the long formula resulting from the complexity analysis of the MINSPAN algorithm. Use the fact that

$$\sum_{i=1}^{n} d_i = 2m$$

3. Imitate the action of the modified MINSPAN algorithm on the graph in Figure 22.1 to find a longest spanning tree.

References

N. Christofides. *Graph Theory: An Algorithmic Approach.* Academic, New York, 1975.

Edward M. Reingold, Jurg Nievergelt, and Narsingh Deo. *Combinatorial Algorithms: Theory and Practice.* Prentice-Hall, Englewood Cliffs, N.J., 1977.

23

GENERATIVE GRAMMARS

Lindenmayer Systems

The growth of certain kinds of plants can be modeled, to an extent, by a formal scheme known as a *Lindenmayer system.* Such systems, first developed by the biologist-mathematician Aristid Lindenmayer in 1968, are really a special kind of generative grammar. In this more general setting, words in a formal language are produced by a stepwise process of replacement. Starting with a single symbol, at each stage one or more symbols in the current word are replaced by certain words given in a list of "productions." When the word has ceased "growing," it is considered to be a member of the language generated by the grammar.

A simple and graphic illustration of this process is provided by the growth of a red alga (Figure 23.1), an example due to Lindenmayer. In this example many intervening steps between each of the three stages shown have been omitted. In fact, starting from an initial "bud" consisting of a single cell, the algal model goes through six cell divisions before even the first illustrated stage is reached. These steps are shown in Figure 23.2.

We return later to Lindenmayer systems and this particular example, but first there are some remarkable properties of generative grammars to discuss.

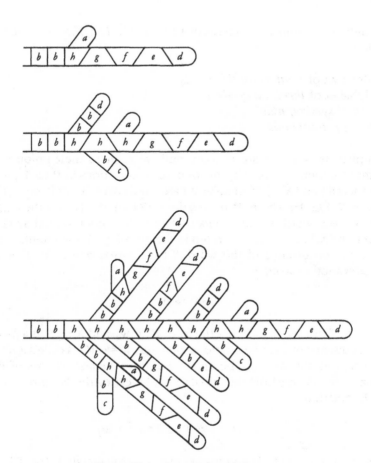

Figure 23.1 Growth of a red alga

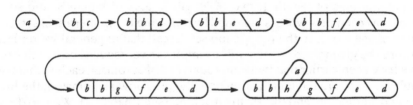

Figure 23.2 The first six steps of growth

153

A generative grammar G consists of a 4-tuple (N, T, n, P) whose elements are as follows:

N: alphabet of *nonterminal symbols*
T: alphabet of *terminal symbols*
n: *initial symbol*, $n \in N$
P: set of *productions*

The alphabets N and T are disjoint, and we refer to their union as A. The grammar G generates words by means of the productions in P. Each production is an ordered pair (X, Y) where X and Y are words over the alphabet A; formally, we write $X, Y \in A^*$, where A^* means the set of all words over the alphabet A. Moreover, the word X must contain at least one nonterminal symbol. The production (X, Y) is normally written in the form $X \rightarrow Y$, the intention being to replace the occurrence of the word X (in a larger word) by Y. A word W *generates* another word W'

$$W \Rightarrow W'$$

if W has the form $W_1 X W_2$, W' has the form $W_1 X' W_2$, and $X \rightarrow X'$ is a production in P. A sequence of such generations, denoted $W \overset{*}{\Rightarrow} W'$, is called a *derivation sequence,* with the last word being *derived* from the first. The set of all words obtainable in this way from the initial symbol is called the *language generated by G.* In notation,

$$L(G) = \{ W : W \in T^*, n \overset{*}{\Rightarrow} W \}$$

For example, the set of all palindromes over the alphabet $\{0, 1\}$ (see Chapter 14) would be produced by the following grammar:

$$N = \{n\}, \quad T = \{0,1\}, \quad n, \quad P = \{n \rightarrow 0n0, \ n \rightarrow 1n1, \ n \rightarrow 0, \ n \rightarrow 1, \ n \rightarrow \lambda\}$$

The simplest illustration of this fact is a derivation tree showing all possible words of a given length resulting from the grammar (Figure 23.3). Thus there are two palindromes of length 1, two of length 2, four of length 3, and so on. Notice how each new word is obtained by replacing the nonterminal symbol n by one of five words in the production set. Recall that in general we replace a subword of a given word and that this example is a very special case of that rule.

We have seen earlier that there are four types of automata, each with its own simple description: the finite automaton, the pushdown automaton, the linear bounded automaton, and the Turing machine (see Chapter 7). We also discovered that there were four language types associated with these automata and

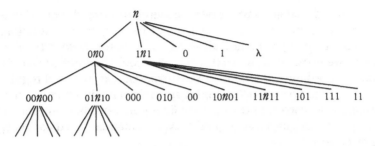

Figure 23.3 The derivation tree for palindromes

that some of these language types have simple descriptions in purely language-theoretic terms. Now we are about to find out that there are simple definitions for four different kinds of generative grammar, each more general than the last, and that these grammars generate precisely the languages accepted by the machines above, respectively!

Let G be a grammar in which each production has the form $x \rightarrow yX$ or $x \rightarrow X$, where $x, y \in N$ and $X \in T^*$. Such a grammar turns out to generate a regular language — which we shall presently prove. In the meantime, we note in passing that the productions used to create palindromes were of a more general type than those employed by the kind of grammar defined above. At the same time, no finite automaton will accept palindromes. (Chapter 14).

Let G be a grammar whose productions all look like $x \rightarrow X$, where $x \in N$ and $X \in A^*$. Because X is a word over alphabet A (the union of terminal and nonterminal symbols), this kind of production includes the production above as a special case. It also includes a palindrome-generating grammar as a special case.

If we now expand the set of allowed production types still further to include those of the form $XyZ \rightarrow XYZ$, where $X, Y, Z \in A^*$ and $y \in N$, we obtain a still more general type of grammar G. The only additional requirement of this grammar is that the word Y, which replaces y, not be empty unless the production has the form $y \rightarrow \lambda$ and y is not in the right-hand word of any production in G.

Note that in this last definition the essential production is $y \rightarrow Y$, but that this replacement takes place within a certain context, namely, $X \ldots Y$. For this reason, G is called a *context-sensitive grammar*. Naturally, the second grammar type defined above, having no context, is called *context-free*.

The languages generated by these grammars are the ones we had called context-free and context-sensitive in an earlier chapter (see Chapter 7). Such languages are accepted by pushdown automata and linear bounded automata, respectively.

The fourth and final grammar to be considered has already been defined. It is the generative grammar in its most general form with no restrictions at all placed on the type of production allowed—except for those in the definition. Since there are no restrictions, such a grammar could be (and is) called a *type 0 grammar*. Accordingly, the context-sensitive, context-free, and regular grammars are also called *type 1, type 2,* and *type 3,* respectively. Once again, the type 0 grammars also correspond to a specific automaton, but only of the most general type; the languages generated by type 0 grammars are those accepted by Turing machines.

We turn now to a demonstration of a grammar/language equivalence in the simplest case, where type 3 grammars generate regular languages, the languages accepted by finite automata.

According to a problem in Chapter 14, if a language L is regular, then so is its *reverse,* the language obtained by reversing all L's words. This fact comes in handy during our demonstration that for every type 3 language (a language generated by a type 3 grammar), there is a finite deterministic automaton which accepts it.

First, let M be a finite deterministic automaton accepting binary inputs. To show that the language L accepted by M is type 3, we construct a grammar G_M as follows: Let $N = \{q_0, q_1, \ldots, q_m\}$, the set of M's states, where q_0 is the initial state. Set $T = \{0, 1\}$, and for each transition

we write $q_i \rightarrow q_j t$, where t is the input symbol and q_i, q_j are states of M. Define P to be the set of all productions obtained in this manner. If we add the transition $q_k \rightarrow \lambda$ for each final state q_k and make q_0 the initial nonterminal symbol, then the grammar G_M generates the reverse of L.

For example if the word 110 is accepted by M, then we may picture a series of transitions like those shown in Figure 23.4. The four productions in Figure 23.4 give rise to the derivation sequence $q_0 \Rightarrow q_1 1 \Rightarrow q_2 11 \Rightarrow q_3 011 \Rightarrow 011$. This is the reverse of 110. In general, if we write a word accepted by M as $a_1 a_2 \cdots a_n$ and if the states entered by M from q_0 to the final state are denoted $q_1, q_2, \cdots, q_n$ (final), then clearly $q_0 \overset{*}{\Rightarrow} a_n \cdots a_2 a_1$ by a simple extension of the argument illustrated above.

If, however, we start with a type 3 language L, then it is easy to construct a finite automaton M_L which accepts the reverse of L. In proving this, it is also convenient to allow M_L to be a nondeterministic automaton which is equivalent for our purposes to a deterministic one (see Chapter 26).

Given a grammar G which generates L, for each production $x \rightarrow yX$ or $x \rightarrow X$ we set up transitions in a corresponding automaton M_L whose states include the

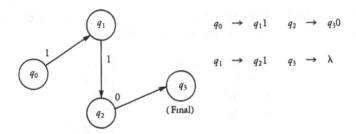

$$q_0 \;\rightarrow\; q_1 1 \qquad q_2 \;\rightarrow\; q_3 0$$

$$q_1 \;\rightarrow\; q_2 1 \qquad q_3 \;\rightarrow\; \lambda$$

Figure 23.4 Turning transitions into productions

nonterminal symbols of G. Additional states are included according to the following recipe: To accommodate a production of the form $x \rightarrow yX$, we not only use x and y as states q_x and q_y, respectively, but also add states which allow the word X to be processed, so to speak. Suppose that $X = x_1 x_2 \cdots x_n$. Then we add $n - 1$ intermediate states between q_x and q_y in the manner shown in Figure 23.5. Transitions other than those prompted by the symbols composing X lead to a single "dead-end" state. Similarly, one encodes a production of the form $x \rightarrow X$ by a string of states, the last one of which is defined to be a final state for M_L. The reason that this construction requires M_L to be nondeterministic is that a given state like q_x in the illustration above may well have several transitions from it which are all triggered by the input symbol x_n (see Chapter 26). In any event, the construction makes clear why M_L accepts words from L and only such words.

The proofs that languages of types 2, 1, and 0 are those accepted by pushdown automata, linear bounded automata, and Turing machines, respectively, are only slightly more difficult than the proof presented above.

Generative grammars are used in the theory and practice of compilers. The words of a programming language, along with the various characters used to define variables, arrays, and other program entities, are themselves symbols in a corresponding language. This is not the programming language as we normally think of it. Rather, each "word" in this language is a syntactically correct program!

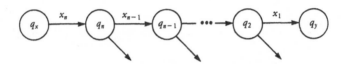

Figure 23.5 Turning productions into transitions

Returning to the Lindenmayer system we started with, we see that the actual productions used to model the growth of the red alga are

$$
\begin{array}{lll}
a \rightarrow b|c & e \rightarrow f & (\rightarrow (\\
b \rightarrow b & f \rightarrow g &) \rightarrow) \\
c \rightarrow b|d & g \rightarrow h(a) & | \rightarrow | \\
d \rightarrow e\backslash d & h \rightarrow h & / \rightarrow \backslash \\
& & \backslash \rightarrow /
\end{array}
$$

The cells and the walls between them are indicated symbolically by small alphabetic letters, vertical and diagonal lines, and parentheses. The vertical line represents a straight wall, the diagonal lines a slanted wall, and the parentheses a new bud. This example has only one defect: The person or device creating a diagram of a resulting stage of growth must remember to alternate the direction of the slanting walls and to draw buds on the resulting long sides of the h cells.

Problems

1. Verify that the red-alga language is context-free and that its grammar is, also.

2. Prove that each type 2 language is accepted by some pushdown automaton. Prove the converse.

3. Add productions to the scheme above which cause the algal h cell walls to alternate their direction. Can you create a grammar which generates images directly by using graphic cells as symbols?

References

A. Salomaa. *Formal Languages.* Academic, New York, 1973.

Gabriel T. Herman and Gregory Rozenberg. *Developmental Systems and Languages.* North Holland, Amsterdam, 1975.

RECURSION

The Sierpinski Curve

The subjects of recursion and fractals go well together. Recursion is the invocation of a computation inside an identical computation that is already in progress. Fractals are shapes that occur inside other, similar shapes. We use a version of the Sierpinski space-filling curve to illustrate recursion in a very concrete way.

The Sierpinski curve is what mathematicians call the *limiting curve* of an infinite sequence of curves numbered by an index $n = 1, 2, 3, \ldots$ A special property of the Sierpinski curve is that if fills two-dimensional space. If each curve in the sequence in Figure 24.1 is drawn to half the scale of its predecessor, then every point in the region of the curve will be found arbitrarily close to some members of the sequence. In other words, the sequence of curves comes closer and closer to every single point in the region. The limiting curve actually covers them all!

Mathematically speaking, a recursive function is one that uses itself in its own definition. Programming languages are said to be recursive if they allow procedures that call themselves. The latter form of recursion, as well as its implementation, is the subject of this chapter.

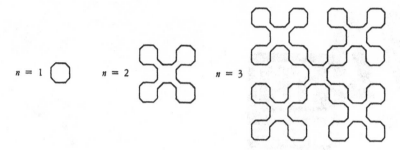

Figure 24.1   The first three curves

To draw the Sierpinski curve, two procedures are used, ZIG and ZAG. Imagine a worm burrowing under the bark of a tree. Whatever direction the worm is currently headed, here's what it must do if asked to ZIG or to ZAG:

ZIG: Turn left, advance distance *d*.
   Turn left again, advance distance *d*.

ZAG: Turn right, advance distance *d*.
   Turn right again, advance distance *d*.
   Turn left, advance distance *d*.

The distance advanced *d* will be variable. Indeed, *d* will be a parameter of the program.

The argument *n* of ZIG and ZAG as defined below will control the recursion process by specifying when it will end.

Both ZIG and ZAG are defined in terms of themselves and each other:

ZIG(n): **if** $n = 1$ **then** *turn* left	ZAG(n): **if** $n = 1$ **then** *turn* right
advance 1 unit	*advance* 1 unit
turn left	*turn* right
advance 1 unit	*advance* 1 unit
else ZIG($n/2$)	*turn* left
ZAG($n/2$)	*advance* 1 unit
ZIG($n/2$)	**else** ZAG($n/2$)
ZAG($n/2$)	ZAG($n/2$)
	ZIG($n/2$)
	ZAG($n/2$)

Having defined both procedures algorithmically, we have only to write the "main program":

ZIG(8)
ZIG(8)

The program consists of two consecutive calls to procedure ZIG with the distance argument equal to 8. Throughout we assume a plotting or screen-drawing facility that is capable of executing the *turn* and *advance* commands used in both procedures.

When ZIG is first called with the argument 8, the procedure first tests whether $n = 1$. Finding that n is not 1, ZIG makes a call to itself with $n = 8/2 = 4$. This call results in another test of the argument against 1. Since the test fails, the new version of ZIG makes yet another call to itself, this time with argument $n = 2$. The argument is not yet 1, but one more call to ZIG changes this situation; when ZIG is executed one more time, the computer executes the drawing command to produce the first small fragment of the fourth Sierpinski curve (Figure 24.2). When the call ZIG(2) was made, it made not one procedure call of its own, but four:

ZIG(1)
ZAG(1)
ZIG(1)
ZAG(1)

We have shown the first ZIG(1) having been executed. The remaining procedure calls add three more segments of the curve to the initial one (Figure 24.3).

The first ZAG actually touches the end of the second ZIG in the diagram. Hereafter we shall show ZIGs and ZAGs with rounded corners to avoid having

Figure 24.2 The execution of ZIG(1)

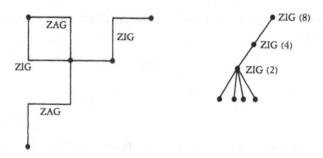

Figure 24.3 The execution of ZIG(2)

the curve intersect itself. The progress of our curve-drawing program can be followed by using a tree: Each execution of ZIG or ZAG either results in some portion of the curve being drawn or results in four more calls to ZIG and ZAG. In the tree diagram above, ZIG(8) resulted in a call to ZIG(4) which resulted in a call to ZIG(2) which resulted in a call to ZIG(1). The last call resulted in no further procedure calls but in the completion of a portion of the curve. This completed the execution of the first ZIG(1) call.

The computer has a means of keeping track of where it is during recursion. In this case, one might think of its memory holding a representation of the ZIG(2) call something like the following:

ZIG(2): ZIG(1)✓
      ZAG(1)
      ZIG(1)
      ZAG(1)

It had completed the first ZIG(1) call. It went on to execute ZAG(1), ZIG(1) again, and ZAG(1) once more to complete the curve drawn above. At such a pass, the execution of ZIG(2) is itself completed, and execution of the program recurses to the next higher level, to ZIG(4). But only the first call within ZIG(4) is thus complete. The next call is to ZAG(2). The latter call results in four more portions of the curve being drawn (Figure 24.4). For extra clarity, the separate ZIGs and ZAGs have been shown as bold lines. By following the program's progress in terms of procedure calls, the tree has now been filled out to the extent of eight terminal nodes. Execution of the ZIG(4) call is now half completed.

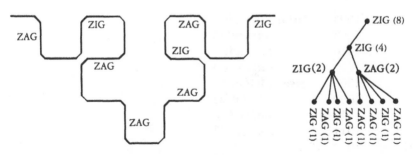

Figure 24.4　ZAG(4) half completed

The rules for executing ZIG(8) (which is, after all, only "half" the curve-drawing program) can be thought of as rules for traversing the tree above. Each time a procedure is called, descend one level of the tree to a new node. Each time a procedure is completed, ascend one level of the procedure that called it in the first place. The final curve can be read off, so to speak, from the various calls to ZIG(1) and ZAG(1) comprising the bottom level of the tree. When the second call to ZIG(8) is complete, the third Sierpinski curve emerges — as in Figure 24.1.

Those who write a Sierpinski curve program in any language having recursion will be rewarded with some very pretty higher-order curves. The highest order possible for a given computer will depend on the resolution of drawing available. For example, with a screen that has at least 256 pixels (picture elements) across its smallest dimension, the eighth-order ($n = 8$) Sierpinski curve may be drawn. Merely execute this program:

ZIG(256)
ZIG(256)

Recursion is controlled in a computer by means of a pushdown stack (see Chapter 7). For the sake of simplicity, we assume that the program, including the definitions of ZIG and ZAG, is stored in consecutive locations starting at memory address 1001:

	1001	ZIG(8)
	1002	ZAG(8)
ZIG:	1003	**if** $n = 1$

	1004	**then** *turn* left
	1005	*advance* 1
	1006	*turn* left
	1007	*advance* 1
	1008	**else** ZIG($n/2$)
	1009	ZAG($n/2$)
	1010	ZIG($n/2$)
	1011	ZAG($n/2$)
ZAG:	1012	**if** $n = 1$

Of course, the program would be stored not in the form shown but in machine language. The essential idea, however, involves storing the address of present and past instructions on a pushdown stack. At first ZIG(8) is executed. The address of this instruction, as well as its associated parameter value $n = 8$, would be placed on the recursion stack. Execution of the program would automatically pass to the instruction at 1003, then (since $n \neq 1$) to 1008. Because the instruction at 1008 is another call to ZIG, the address of this instruction and its current parameter value, $n/2 = 4$, are pushed onto the stack which now has the appearance shown below:

1008	4
1001	8

The call to ZIG with parameter value 4 results in another call with parameter value 2 and, finally, a call to ZIG with value 1:

1008	1
1008	2
1008	4
1001	8

At this stage, however, $n = 1$ and the instructions at addresses 1004 to 1007 are executed, completing the execution of ZIG(1). When recursion returns one level, the stack is popped:

1008	2
1008	4
1008	8

Execution now passes to the next instruction, the one at 1009. The stack then contains

1009	1
1008	2
1008	4
1008	8

In this way the stack contents are alternately pushed and popped as the instructions at 1008, 1009, 1010, and 1011 are executed with $n = 1$. When the last of these, ZAG(1), is complete, execution returns from the procedure ZIG(2) called from line 1008, then passes to the procedure ZAG(2) called from line 1009.

It is not difficult to see that the number of addresses occupying the recursion stack at any given time is simply the depth in the recursion tree to which the procedure has descended. Thus is recursion implemented, at least in outline.

It seems worth mentioning that the definition of the Sierpinski curve can be given in terms of a context-free grammar (see Chapter 23). If, for example, we use letters such as *a* and *b* to represent ZIG and ZAG, respectively, drawn, then the Sierpinski curve drawing process can be represented as follows:

aa
a → *abab*
b → *bbab*

Here, *aa* represents an initial word. Alternately replacing *a* by *abab* and *b* by *bbab* results in ever-lengthening words:

aa
abababab
ababbbabababababbbabababbbabababbbab
.
.
.

At any point one may stop, replacing each letter by its corresponding curve segment, appropriately oriented. The result for the *n*th word in the sequence will be the *n*th Sierpinski curve.

Problems

1. By redefining the drawing steps in the **if** condition of ZIG and ZAG, it is possible to obtain Sierpinski curves with cut corners like the ones shown in Figure 24.4. Find such a definition.

2. Write a recursive procedure for the tower of Hanoi problem (see Chapter 55) by assuming the existence of an instruction called *transfer*(a,b) that transfers a disk from peg a to peg b.

3. A recursive program can always be replaced by a purely iterative, albeit much longer, one. Show how you might do this in general by making the recursion stack "visible," i.e., by making it part of the equivalent iterative program.

References

Kurt Maly and Allen R. Hausar. *Fundamentals of the Computing Sciences.* Prentice-Hall, Englewood Cliffs, N.J., 1978.

Niklaus Wirth. *Algorithms + Datastructures = Programs.* Prentice-Hall, Englewood Cliffs, N.J., 1976.

25

FAST MULTIPLICATION

Divide and Conquer

t would be interesting to know the number of multiplications performed daily by computers worldwide. That figure is probably somewhere between 10^{15} and 10^{20} currently. In most computers, multiplication absorbs only a few microseconds. Multiplication circuits are already optimized to produce the product of two, say, 32-bit numbers in the shortest possible time. However, the availability of fast hardware solutions to the problem of multiplying two 32-bit numbers tends to obscure a very general question and one which may have more than just theoretical importance: How fast can two n-bit numbers be multiplied? Rather than attempt a generalized hardware scheme for fast multiplication, we will instead assume that all operations are carried out at the bit level and then merely attempt to determine the smallest number of bit operations necessary to the formation of products.

The bit context of this problem can be illustrated by considering binary addition for a moment:

$$
\begin{array}{r}
1\ 0\ 0\ 1 \\
+\ 1\ 1\ 0\ 1 \\
\hline
1\ 0\ 1\ 1\ 0
\end{array}
$$

$$
\begin{array}{cccc}
1 & 0 & 0 & 1 \\
+ & + & + & + \\
1 & 1 & 0 & 1 \\
\downarrow & \downarrow & \downarrow & \downarrow \\
1 \leftarrow 0 & 1 & 1 \leftarrow 0
\end{array}
$$

In adding the two numbers 1001 and 1101, the bit operations are represented by the columns in the figure. Starting at the right-hand end, we add 1 and 1 to obtain 0 and propagate a carry to the next column. This may be considered as all one bit operation. Clearly, no matter how large n is, this operation may be taken as fundamental and absorbs, say, 1 unit of time. Equally clearly, to add two n-digit binary numbers requires n such bit operations.

In the case of multiplication, however, the situation is very different. First, by adapting the ordinary rules for multiplication learned in elementary school, it is not hard to see that two n-bit numbers can be multiplied in $n^2 + 2n - 1$ units of time.

$$
\begin{array}{r}
1\ 0\ 0\ 1 \\
\times\ 1\ 1\ 0\ 1 \\
\hline
1\ 0\ 0\ 1 \\
0\ 0\ 0\ 0 \\
1\ 0\ 0\ 1 \\
1\ 0\ 0\ 1 \\
\hline
1\ 1\ 1\ 0\ 1\ 0\ 1
\end{array}
$$

It requires n^2 steps to form n intermediate products and then another $2n - 1$ steps to add them. In practice, we ignore the $2n - 1$ term and concentrate only on the n^2 steps, saying that this sort of multiplication requires "in the order of" n^2 units of time (see Chapter 15), or

$$O(n^2) \text{ steps}$$

It would be our primary aim to reduce the power of n in this expression. Can we get it down to $O(n^1)$, for example?

As an initial attempt at speeding up bit-level multiplication, let us take the divide-and-conquer approach: suppose that two n-bit numbers x and y are to be multiplied and that each number is split into two equal-length parts as follows:

$$x = a \times 2^{n/2} + b \quad \text{and} \quad y = c \times 2^{n/2} + d$$

The splitting process is carried out by scanning halfway along each number and merely splitting it in two. (If there is no "halfway," one can easily be created by adding an additional, high-order 0 bit to the number.) For example,

$$1001 = 10 \times 2^2 + 01 \quad \text{and} \quad 1101 = 11 \times 2^2 + 01$$

Having split our numbers x and y into two parts each, we can now rewrite one multiplication as four separate ones:

$$x \times y = (a \times c) \times 2^n + (b \times c) \times 2^{n/2} + (a \times d) \times 2^{n/2} + b \times d$$

At first glance, it appears that four multiplications of $n/2$-bit numbers are required and that $4(n/2)^2 = n^2$ multiplications will be necessary. This is hardly a promising start!

But if $a \times c$ and $b \times d$ have already been computed, then it is not necessary to carry out two multiplications to form the products $b \times c$ and $a \times d$. In fact,

$$b \times c + a \times d = (a + b) \times (c + d) - a \times c - b \times d$$

How many bit operations will be required to carry out the various processes implicit in these observations? To form a and b from x requires $n/2$ bit operations in each case, and we assume, for the time being, that products $a \times c$ and $b \times d$ are formed by the "elementary school method" and thus require $(n/2)^2 + 2(n/2) - 1$ bit operations each. The sums $a + b$ and $c + d$ each require $n/2$ bit operations, since each has $n/2$ bits, and the product of these quantities needs $(n/2 + 1)^2 + n + 1$ such steps. Summarizing this information in tabular form, we obtain the following:

Process	Number of steps	Number of bits in result
a and b	$n/2$ each	$n/2$ each
$a \times c$ and $b \times d$	$(n/2)^2 + 2(n/2) - 1$ each	n each
$a + b$ and $c + d$	$n/2$ each	$n/2 + 1$ each
$(a + b) \times (c + d)$	$(n/2 + 1)^2 + 2(n/2 + 1) - 1$	$n + 2$
$(a + b) \times (c + d) - a \times c - b \times d$	$2n$	$n + 2$

169

In the last row of the table, the product $(a + b) \times (c + d)$ is already available as an $(n + 2)$-bit number while $a \times c$ and $b \times d$ are available in n-bit form, so that adding the (negative) second and third terms to the first requires a total of $2n$ bit operations, not counting the propagation of carrys. Adding the total number of steps taken in forming the final result, we obtain

$$3\left(\frac{n}{2}\right)^2 + 15\left(\frac{n}{2}\right) \text{ bit operations}$$

This quantity is roughly $3n^2/4$, however, and unless we had the creative idea of using the same technique again for the computation of $a \times c$, $b \times d$, and $(a + b) \times (c + d)$, we would be stuck with only a constant factor of improvement. Indeed, in this as in many other problems (see Chapter 32), we must divide again and again before we have truly conquered!

Let $T(n)$ denote the number of bit operations achievable in this way in the multiplication of two n-bit numbers. By simply inserting $T(n/2)$ and $T(n/2 + 1)$ in the appropriate places in the table above, we obtain the following recursion formula:

$$T(n) = 2T\left(\frac{n}{2}\right) + T\left(\frac{n}{2} + 1\right) + 8\left(\frac{n}{2}\right)$$

$$= 3T\left(\frac{n}{2}\right) + cn \qquad \text{where } c \text{ is an appropriately chosen constant}$$

It is now easy to show by the method of induction that

$$T(n) \leq 3cn^{\log 3} - 2cn$$

This upper bound on the speed with which two n-digit numbers can be multiplied may be re-written as $O(n^{1.59})$. This is not yet down to $O(n^1)$, but it is a real improvement over $O(n^2)$.

The fast multiplication technique just described was published by A. Karatsuba and Y. Ofman in 1962. It remained the best result of its kind until it was superceded in 1971 by a new technique discovered by A. Schönhage and V. Strassen. These workers developed a new divide-and-conquer algorithm requiring only

$$O(n \log n \log\log n) \text{ bit operations}$$

an improvement which comes very close to our (perhaps unachievable) goal of $O(n^1)$!

We turn now to the problem of multiplying two $n \times n$ matrices. In the analysis which follows, we no longer examine operations at the bit level, since n is no longer the number of bits but the size of the matrix. In adopting such a rationale, we assume a computer which carries out multiplications just as quickly as additions. This is not unreasonable since

Most computers have high-speed parallel circuitry for multiplication.

The matrix entries will be assumed always to fit within the word size of whichever computer we happen to be using.

No matter what constant factor separates the time of a multiplication from the time of an addition, the same order-of-magnitude figure (as a function of n) will be obtained.

The elementary method of multiplying two matrices X and Y is to use the definition directly: The ijth element of the product will be

$$\sum_{k=1}^{n} x_{ik} y_{kj}$$

and clearly $O(n^3)$ operations (additions and multiplications) will be required, $O(n)$ operations being used to generate each of n^2 entries.

At the same time, since the product $Z = X \times Y$ has n^2 entries, we cannot expect to get away with fewer than $O(n^2)$ operations. In what follows, we assume that n is a power of 2. If it was not, we could nevertheless pad the matrices with 0s so that it was — the product Z would still be found lying intact within the product of such inflated matrices.

As a first step, we partition X, Y, and Z into $n/2 \times n/2$ submatrices as follows:

$$\begin{pmatrix} Z_{11} & Z_{12} \\ Z_{21} & Z_{22} \end{pmatrix} = \begin{pmatrix} X_{11} & X_{12} \\ X_{21} & X_{22} \end{pmatrix} \times \begin{pmatrix} Y_{11} & Y_{12} \\ Y_{21} & Y_{22} \end{pmatrix}$$

Simple algebraic manipulation shows that if we form the intermediate matrices

$$W_1 = (X_{11} + X_{22}) \times (Y_{11} + Y_{22})$$

$$W_2 = (X_{21} + X_{22}) \times (Y_{11})$$

$$W_3 = (X_{11}) \times (Y_{12} + Y_{22})$$

$$W_4 = (X_{22}) \times (Y_{11} + Y_{21})$$

$$W_5 = (X_{11} + X_{12}) \times (Y_{22})$$

$$W_6 = (X_{21} - X_{11}) \times (Y_{11} + Y_{12})$$

$$W_7 = (X_{12} - X_{22}) \times (Y_{21} + Y_{22})$$

171

then the four submatrices of Z are easily obtained as sums of W matrices:

$$Z_{11} = W_1 + W_4 - W_5 + W_7 \qquad Z_{21} = W_2 + W_4$$
$$Z_{12} = W_3 + W_5 \qquad\qquad Z_{22} = W_1 + W_3 - W_2 + W_6$$

If we computed the submatrices Z_{ij} directly as matrix products among the X_{ij} and Y_{ij}, then eight matrix multiplications would be needed. The method above requires only seven, and armed with this knowledge, we let $T(n)$ denote the total number of operations required by a continued divide-and-conquer strategy. The resulting recurrence formula for $T(n)$ becomes

$$T(n) = 7T\left(\frac{n}{2}\right) + 18\left(\frac{n}{2}\right)^2$$

Using induction and the fact that $T(1) = 1$, we can show that this recurrence has the solution

$$T(n) = 7^{\log n} + 18n^2 \sum_{k=0}^{\lfloor \log n \rfloor - 1} \left(\frac{7}{4}\right)^k$$
$$\doteq 7n^{\log 7}$$
$$= O(n^{2.81}) \text{ operations}$$

In the case of matrix multiplication, there has not been the same dramatic improvement as occurred in the case of integer multiplication. The $O(n^{2.81})$ technique described above was developed by V. Strassen in 1969. Two further improvements followed in quick succession in 1979. First, A. Schönhage and then V. Pan found methods requiring $O(n^{2.73})$ and $O(n^{2.61})$ operations, respectively. Is this last order of magnitude the best possible? One hears rumors that an $O(n^{2.55})$ algorithm may be achievable.

It is notoriously difficult to analyze problems with a view to establishing lower bounds for the worst-case efficiency of any algorithm that might solve them. A few problems, however, have notoriously easy solutions in this respect. For example, to search an unordered list of n items will always take $O(n)$ steps in the worst case, no matter what algorithm one proposes. It is equally easy to show that to multiply two $n \times n$ matrices will require at least $O(n^2)$ steps since each matrix has this many entries and each entry will have to be considered at least once by any algorithm that carries out the multiplication.

This leaves theoreticians with the perplexing question, "How much higher than n^2 is the true lower bound for the complexity of the matrix multiplication problem?

Problems

1. The expression for $T(n)$, the number of bit operations to perform bitwise multiplication of two n-bit numbers, assumes that n is a power of 2. Using this assumption, prove the bound on $T(n)$ given in this chapter. If n is not a power of 2, how would you alter this expression?

2. The fast multiplication technique of Schönhage and Strassen requires $O(n \log n \log\log n)$ bit operations. Show for large enough n that this quantity is smaller than $O(n^{1+\epsilon})$, where ϵ is an arbitrarily small but fixed positive real number.

3. Would dividing either integers or matrices into three parts instead of two yield superior results? Imitate either argument, and solve the resulting recursion to answer this question.

References

J. E. Savage. *The Complexity of Computing*. Wiley-Interscience, New York, 1976.

Alfred V. Aho, John E. Hopcroft, and Jeffrey D. Ullman. *The Design and Analysis of Computer Algorithms*. Addison-Wesley, Reading, Mass., 1974.

NONDETERMINISM

Automata that Guess Correctly

I t is night, and you are driving through a strange city. You must be at X's house in 10 minutes, but you have no address. You cannot even phone X because you do not know her last name. Luckily, your car is equipped with a nondeterministic automaton built into the instrument panel. Touch a red button on its impassive-looking case, and one of five letters instantly appears in a display.

L F R S B

Here L means left, F means forward, R means right, S means stop, and B means backward. Each time you come to an intersection, you push the red button and follow the direction given. Eventually, you reach an intersection for which the letter B appears: You have overshot X's house. Now you must turn around and touch the red button as you pass each house. Eventually an S appears. You enter that house to find your friend waiting for you.

No such automata exist in reality, but they are very useful in theory. Not only do they result in a certain simplification of the theory of automata, but also they lead to one of the most profound open questions in computer science (see Chapter 45).

Each of the automata in the Chomsky hierarchy (see Chapter 7) has a deterministic and nondeterministic version. In this chapter we study nondeterministic models for all four types of automata. The first is the finite automaton whose deterministic version we studied in Chapter 2.

In the state-transition diagram (Figure 26.1) we may discover a multitude of routes from the initial state I to the final or accepting state F. However, if we ask what input words put this automaton in its final state, a certain confusion results: Some states have two or more transitions which are triggered by the same input symbol. Indeed, an input word like 01011 can put the machine in any one of several possible states, one of which is the final, or accepting, state. The definition of a nondeterministic finite automaton says that such a word is accepted by the machine: If the automaton *could* accept a given word, then it is considered that the automaton *does* accept the word.

In precise terms, a nondeterministic finite automaton (NFA) consists of a finite set Q of states, a finite alphabet Σ, and a transition function δ. The same thing is true of a deterministic automaton. Only the transition function is different:

$$\delta: Q \times \Sigma \rightarrow P(Q) - \varnothing$$

where $P(Q)$ represents the power set of Q, namely, the set of all subsets of Q. In other words, when the NFA is in a given state and a given symbol is inputted, then a nonempty subset of next states results. In the automaton displayed below, the transition function δ maps the state-input pair $(I,0)$ into the state set $\{A,B\}$.

If w is a word composed of letters from Σ, we define $\delta(w)$ to be the set of all states that an NFA might be in by the time the last symbol of w has been inputted

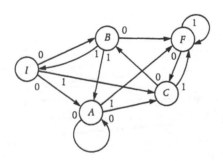

Figure 26.1 A nondeterministic automaton

to it. If $\delta(w)$ contains a final state, then w is *accepted* by the NFA. The set of all words accepted by a given NFA is called its *language*.

It is worth stressing, once again, that this definition of acceptance is tantamount to saying that the automaton guesses the right transition at each stage of processing. For example, the input word 01011 could result in one series of transitions such as $I \rightarrow B \rightarrow I \rightarrow B \rightarrow I \rightarrow C$ or in another series such as $I \rightarrow A \rightarrow F \rightarrow C \rightarrow F \rightarrow F$. As far as determining the acceptance of 01011 by the automaton, however, only a sequence like the latter is assumed to be computed.

One might suppose that a finite automaton with such extensive guessing powers would accept languages well beyond the powers of its deterministic colleagues. Interestingly, this is not the case at all!

Given an NFA, say M, we may describe a deterministic automaton $\overline{M}$ which accepts exactly the same language as M. Here is how it works.

Suppose M has state set Q, alphabet Σ, and transition function δ. Now let $\overline{M}$ be the automaton with one state q_i for each nonempty subset Q_i of Q. Thus, in writing a state of $\overline{M}$ as q_i, we are referring also to a particular nonempty subset Q_i of Q. The reason for this rather exotic construction lies in a very simple observation: Even if we cannot determine what state M will be in after a given input, we can certainly determine the set of states that M might *possibly* be in.

If M might be in any state of the subset Q_i at a given moment and if a symbol σ is inputted, then following this, M might be in any state of the set:

$$\overline{\delta}(Q_i, \sigma) = \bigcup_{q \in Q_i} (q, \sigma)$$

The function $\overline{\delta}$ is actually the transition function for $\overline{M}$. By denoting $\overline{\delta}(Q_i, \sigma)$ by Q_j, a multitude of transitions in M is expressed as a single transition in $\overline{M}$, as in Figure 26.2. Given this definition of $\overline{M}$'s transition function $\overline{\delta}$, it is clear that $\overline{M}$ is

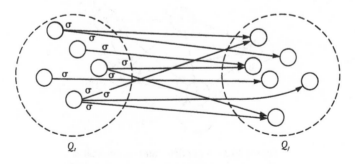

Figure 26.2 Many transitions become one.

deterministic: For each current state and input, there is exactly one new state possible. If q_0 is the initial state of M and F is the set of final states, then we define $Q_0 = \{q_0\}$ to be the initial state of $\overline{M}$. Moreover, any state Q_k of $\overline{M}$ having the property that $Q_k \cap F$ is nonempty qualifies as a final state of M.

To show that $\overline{M}$ accepts L, the language accepted by M, let w be an arbitrary word in L. If w is composed of the symbols

$$w_1 w_2 \cdots w_n$$

then we observe that $\overline{M}$ passes through a succession of states

$$Q_1, Q_2, \cdots, Q_n$$

where

$$\overline{\delta}(Q_{i-1}, w_i) = Q_i \qquad i = 1, 2, \ldots, n$$

Now $Q_1, Q_2, \cdots, Q_n$ is also a succession of subsets of Q, the states of M. It is easily seen that

$$Q_1 = \delta(q_0, w_1) \qquad Q_2 = \delta(q_0, w_1 w_2) \qquad \cdots \qquad Q_n = \delta(q_0, w_1 w_2 \cdots w_n)$$

In other words, as the word w is inputted to M, it may be in any of the states of Q_i when the ith symbol is read. Since w is accepted by M, Q_n contains a final state of M and therefore Q_n is a final state of M. The reverse implication also holds, if w is accepted by $\overline{M}$, then w is also accepted by M. Thus $\overline{M}$ accepts the same language as M, and the result is proved.

Since every deterministic finite automaton (DFA) is just a special kind of NFA, it follows that every language accepted by a DFA is also accepted by an NFA. We have now shown that NFAs are not more powerful than DFAs when it comes to accepting languages. We note in passing, however, that to obtain a DFA equivalent to a given NFA, we had to do an exponential amount of work in constructing it. In general, the equivalent DFA can have as many as 2^n states when the NFA has just n.

At the next level of the Chomsky hierarchy, we discover an automaton whose deterministic and nondeterministic versions are not at all equivalent. Nondeterministic pushdown automata accept context-free languages (see Chapter 7), but some of these languages are not accepted by deterministic pushdown automata.

For example, it is easy enough to show that the following language is accepted by some nondeterministic pushdown automaton:

$$L = \{0^a 1^b 0^c 1^d : a = c \text{ or } b = d\}$$

It is not so easy to show that no deterministic pushdown automaton can accept L. Here we can at least argue intuitively why this is so: In processing the string of symbols $0^a1^b0^c1^d$, the pushdown automaton must somehow "remember" both a and b by storing information about these numbers on its stack. The information might consist of the 0 and 1 strings themselves or something more sophisticated. In any case, merely accessing the information about a (in order to compare a with c) results in a loss of information about b. That information was added to the stack *after* the information about a and must have been popped (i.e., destroyed) in order for the pushdown automaton to recall a.

It follows from this argument (once made rigorous) that nondeterministic pushdown automata are definitely more powerful than deterministic ones. What is the situation with the next higher machines on the Chomsky hierarchy, linear bounded automata?

These devices resemble Turing machines in all respects but one: The amount of tape they are allowed to use is bounded by a linear function of the size of the input string. In other words, each linear bounded automaton has associated with it some constant k, so that when a word of size n appears as input on its tape, the machine is allowed the use of no more than kn tape squares to determine the word's acceptability. It has been a problem of long standing to decide whether nondeterministic linear bounded automata are more powerful than deterministic ones. No one knows.

At the top of the Chomsky hierarchy reside the Turing machines. For these abstract devices, as for finite automata, nondeterminism adds nothing to their language-accepting ability. Given a nondeterministic Turing machine M, an equivalent deterministic machine M' is constructed in the manner of a universal Turing machine (see Chapter 51).

The transition function of M is stored as a sequence of quintuples on a special work tape of M'. To simplify the argument, assume that M has just one nondeterministic transition and that this transition involves just two alternatives that we label 0 and 1:

$$(q, s) \quad\Longrightarrow\quad \begin{matrix} \longrightarrow & q_1, s_1, d_1 & (0) \\ \longrightarrow & q_2, s_2, d_2 & (1) \end{matrix}$$

As in the case of a universal machine, M' scans the tape of M to see what symbol M is reading. In one portion of its work tape M' has written M's current state. It is therefore a simple matter (although complicated to specify in detail) for M' to run through the list of M's quintuples to see which one applies to the current input symbol and state combination. If there is only one, all is well: M' merely writes the symbol to be written in its version of M's tape, changes the state written in its state memory section, and then moves M's (simulated) read/write head in the appropriate direction. But what happens if *two* such transitions are possible?

Here is where M' differs from the universal machine. In one portion of its work tape, it keeps a count that starts at binary 0 and progresses ad infinitum if necessary. The number stored in the counting space tells M' which of the two nondeterministic transitions M will take. Each time M' encounters the state-symbol pair (q,s) in its simulation of M's activity, it consults the current count. If it is consulting this number for the first time, it merely uses the first digit (0 or 1) to decide between the alternatives. If consulting the count for the second time, it uses the second digit, and so on. Sooner or later it either accepts the word inputted to M or consults the last digit in the number. Acceptance by M' implies acceptance by M. But if M' reaches the last digit of the count without accepting M's input word, it adds 1 to the count and begins all over again to simulate M. If M never halts, neither, obviously, does M'. But acceptance by M' is tantamount to acceptance by M. The converses of both of these statements are also true.

The point of M' keeping count is that sooner or later the digits of this slowly and steadily incrementing number must match an accepting sequence of choices for the two (q,s) options by M.

Throughout this chapter we have allowed nondeterminism to reside solely in the choice of transitions which an automaton may make. In the case of Turing machines, there is a second way: While leaving all transitions deterministic, one may assume the appearance on a nondeterministic Turing machine's tape of a finite word x in a position immediately adjacent to the input word w. If, instead of being allowed to choose between two or more transitions for a given state-symbol combination, the Turing machine is directed to consult the word x as a part of its decision-making process, then a machine equivalent to the first type may always be constructed.

This particular notion of nondeterminism is important in defining the all-important notion of nondeterministic polynomial-time complexity (see Chapter 45).

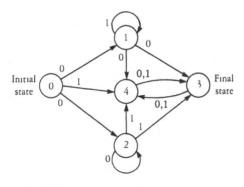

Figure 26.3 What does an equivalent deterministic automaton look like?

Problems

1. Given the NFA in Figure 26.3, use the construction of the equivalence theorem to build a DFA which accepts the same language.

2. Design a nondeterministic pushdown automaton which accepts the language $L = \{0^a1^b0^c1^d: a = c \text{ or } b = d\}$ by final state. This can be done by invoking exactly one nondeterministic transition.

3. A deterministic Turing machine was enabled to simulate a Turing machine having one nondeterministic transition by the simple expedient of counting in binary. Expand this scheme to handle any amount of nondeterminism.

References

H. R. Lewis and C. H. Papadimitriou. *Elements of the Theory of Computation.* Prentice-Hall, Englewood Cliffs, N.J., 1981.

Daniel I. A. Cohen. *Introduction to Computer Theory* John Wiley, New York, 1986.

27

PERCEPTRONS

A Lack of Vision

N eurons, in both humans and animals, are delicate, highly complex cells which are somehow involved in thinking and making decisions (Figure 27.1). The neuron's cell membrane is capable of sustaining an electric charge. When this charge reaches a certain threshold, the neuron "fires." A wave of depolarization spreads rapidly over the cell's surface and travels along its outgoing axon in the form of a nerve impulse, a wave of rapidly increasing (followed by rapidly decreasing) charge. Even though the axon splits into a *dendritic tree,* the nerve impulse travels each branch of the tree, arriving ultimately at a small bulb adjacent to some other neuron with which it communicates by a synapse. If the synapse is excitatory, the impulse arriving at the second cell will increase its surface charge and may even cause it to fire. But if the synapse is inhibitory, the cell is usually prohibited from firing for a time.

These basic features of human and animal neurophysiology were already well known in 1941 when Warren McCullough, a mathematician and doctor, and Walter Pitts, a neurophysiologist, decided to construct a simple model of neurons and their interconnections. About the only property of real systems reflected with reasonable accuracy in their model was the all-or-nothing char-

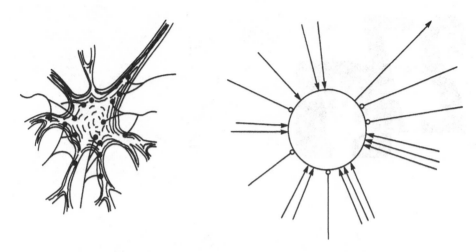

Figure 27.1 A "real" neuron and its formalized version

acter of how real neurons fire. The model was inspired by McCullough's vision of the logical quality of such electrical activity: "If neurons A, B, and C fire, then so will E, unless D fires" translated to "If propositions A, B, and C are true and if D is false, then E is true." McCullough and Pitts proved that any proposition in logic (see Chapter 13) could be realized by a network of their neurons. This result was believed by some people at the time to constitute a primitive explanation of how people actually think. It formed a cornerstone of what might be called the *cybernetic age,* a period of several decades in which many scientists believed that artificial brains were just around the corner, a time in which vain imaginings far outstripped actual results.

A *perceptron* is a special sort of neural net which examines and classifies visual patterns presented to it on a gridlike retina. It does this by weighing evidence submitted to it by a number of component devices, each of which looks only at a specified portion of the retina.

For example, a perceptron Π which classifies patterns into two sets (Figure 27.2)

R: collections of disjoint rectangles

$\bar{R}$: patterns not in R

may easily be constructed. The perceptron Π is visualized as a central evidence-weighing unit receiving inputs from its component devices deployed over the retina as in Figure 27.3.

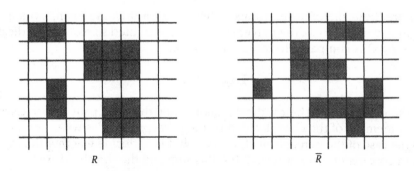

Figure 27.2 Rectangles and nonrectangles

The component devices which gather evidence about the current retinal pattern x are called *predicates* because

Each one examines a finite number of cells in the retina, called its *support*. Each one is a logical function of its support variables. Each cell under its purview may be regarded as a boolean variable which is 1 if that cell is dark and 0 otherwise.

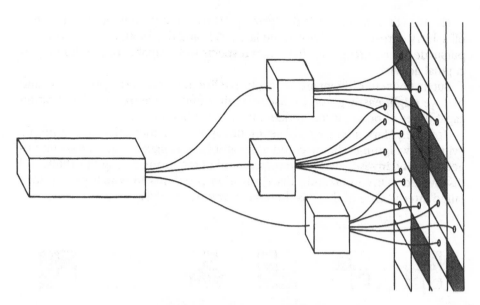

Figure 27.3 A perceptron in action

183

Each predicate therefore transmits either a 1 or a 0 to the main, evidence-weighing component of the perceptron. Here a weighted sum of the predicate values $\phi_i(x)$ is compared to some threshold θ:

$$\Sigma\alpha_i\phi_i(x)\colon \theta$$

If the sum does not fall below the threshold, then the input pattern is declared to be a member of the class decided by the perceptron: otherwise, not.

In the case of the perceptron Π, whose job it is to decide whether the input pattern consists of disjoint rectangles, this sum and the threshold are

$$\Sigma(-1)\phi_i(x)\colon 0$$

The latter is computed very simply. For the ith intersection point on the retina, there are four tangent cells and a predicate ϕ_i which examines these four alone. It generates a 1 if any of the following six patterns are found in these cells (Figure 27.4). It should not be difficult to realize that a pattern presented to Π is a disjoint collection of rectangles if (and only if) none of the six subpatterns above appear anywhere in it. But none of these will appear if (and only if) each $\phi_i(x) = 0$. This last condition is equivalent to writing

$$\Sigma\phi_i(x)\leq 0 \qquad \text{or} \qquad \Sigma(-1)\phi_i(x)\geq 0$$

The idea of perceptrons is due principally to Frank Rosenblatt who touched off a flurry of research activity in the late 1950s and the 1960s by those who felt pursuaded that perceptrons in certain respects were capable of intelligent perceptual activity.

But just how intelligent are perceptrons? Encouraged by the previous example, we might easily go on to discover other impressive pattern classification capabilities of perceptrons. Indeed, many have been found.

However, Marvin Minsky and Seymour Papert, convinced that such a simple evidence-weighing scheme could not possibly be capable of genuinely intelligent activity, investigated the *limitations* of perceptrons during the 1960s and discovered many interesting discrimination problems in which perceptrons failed utterly. One of these is the *connectedness problem*.

Figure 27.4 Local patterns that produce nonrectangles

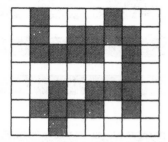 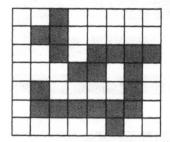

Figure 27.5 Connected (left) and nonconnected (right) patterns

Shown in Figure 27.5 are two patterns. The one on the left is connected, but the one on the right is not. Can a perceptron which says yes every time it is presented with a connected pattern be built? The answer to this question depends on how we pose it. If, for example, the perceptron has a fixed $n \times n$ retina, then one may construct a single predicate capable of distinguishing connected patterns from nonconnected ones. In logical terms, one could construct a massive truth table with 2^{n^2} entries, one for each possible pattern. The truth value of the predicate would be 1 for each table entry corresponding to a connected pattern.

Such a construction, however, runs counter to the spirit of perceptron research in which the weighted sum certainly plays a central role. The only way to prevent the study from degenerating into single-predicate schemes such as the foregoing is to place some limitations on the predicates themselves. Two such limitations studied by Minsky and Papert are quite reasonable from a practical point of view:

Diameter-limited perceptrons: Each predicate receives inputs from within a square k units on a side.

Order-limited perceptrons: Each predicate receives inputs from at most k cells of the retina.

To prevent the perceptron from taking unfair advantage of these limitations by choosing large values of k, we also make the retina infinite. This, of course, implies that we must allow the perceptron to have an infinite number of predicates.

As a sample of some of the interesting and elegant constructions of Minsky and Papert, we now show why no diameter-limited perceptron is capable of always discriminating a connected pattern from a nonconnected one.

Consider the four patterns in Figure 27.6, each one more than k cells long. Let Γ be a diameter-limited perceptron whose designer claims that it can always tell

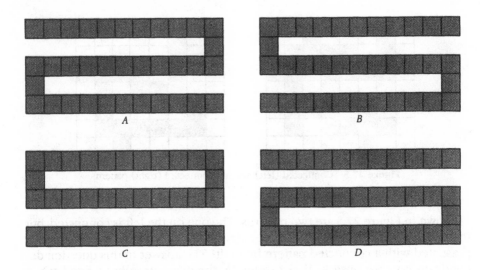

Figure 27.6 Fooling a diameter-limited perceptron

a connected pattern from a nonconnected one. We present Γ with pattern C, and it answers no. Note that each of Γ's predicates falls into one of three simple classes:

L: those which examine at least one cell on the extreme left of the pattern
R: those which examine at least one cell on the extreme right of the pattern
S: those not in L or R

Thus the threshold inequality, currently unsatisfied, may be written

$$\sum_{l \in L} \alpha_l \phi_l(C) + \sum_{s \in S} \alpha_s \phi_s(C) + \sum_{r \in R} \alpha_r \phi_r(C) < \theta$$

If we now replace pattern C by pattern A, only the left-hand end of the pattern changes, so that only the sum over L may change (the other two sums remain unchanged). Since Γ must answer yes for A, the inequality is satisfied and it must be true that

$$\sum_{l \in L} \alpha_l \phi_l(A) > \sum_{l \in L} \alpha_l \phi_l(C)$$

Similarly, in replacing C by B, only the right-hand end of the pattern changes, and we have

$$\sum_{r \in R} \alpha_r \phi_r(B) > \sum_{r \in R} \alpha_r \phi_r(C)$$

The inequality is once again satisfied since B is connected. Now either increase is enough to drive the sum up over the threshold θ. What happens when we present Γ with D? Here, *both* sums have increased, with the S sum remaining unchanged, and the inequality is certainly satisfied, whence the perceptron Γ must reply yes to D which, unfortunately, is not connected.

Minsky and Papert went on to show that no order-limited perceptron may recognize connected figures. There are numerous other tasks at which perceptrons fail, and this work serves as a warning to those who seek machine models of human intelligence that simplicity may not be a feature of the schemes by which we see, think, and act!

Problems

1. Design a predicate that permits a perceptron to distinguish any and all patterns consisting of vertical stripes. Use only 2×2 squares.

2. Devise an argument that shows order-limited perceptrons are also incapable of recognizing connected figures.

References

M. Minsky and S. Papert. *Perceptrons: An Introduction to Computational Geometry.* M.I.T. Press, Cambridge, Mass., 1969.

F. Rosenblatt. *Principles of Neurodynamics.* Spartan Books, New York, 1962.

28

ENCODERS AND MULTIPLEXERS

Manipulating Memory

There is a class of logic circuits concerned mainly with the manipulation of information flow within computers. This class includes encoders, multiplexers, and related devices. Before we introduce them, we provide a single example of their usefulness.

We may think of a computer's memory as consisting of thousands of registers, each register (or "word") consisting of many bits. Consider, for example, a very simplified memory of 4-bit registers as shown below. This memory consists of eight words, a memory address register (MAR), and a memory buffer register (MBR), as shown in Figure 28.1.

When a specific word of information is to be stored in a computer's memory, it is placed in the MBR prior to being stored. Meanwhile, the MAR contains the address in memory of where this word will ultimately be located.

The MBR is connected by input lines to each word of memory; the only

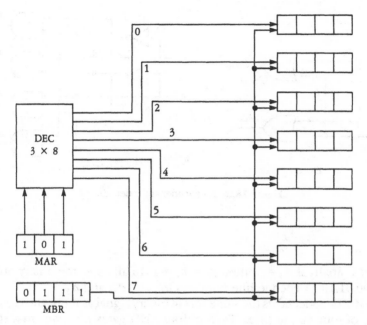

Figure 28.1 A simple computer memory

requirement for storing the MBR contents (say, 0111) at a given address is to signal the register at that address to load the information already available to it from the MBR. The signal to load the MBR contents comes from the MAR via one of the devices studied in this chapter, namely, a decoder (DEC). It takes the three input lines from the MAR and decodes them; each of the eight output lines corresponds to a particular bit pattern in the MAR. Thus, if the MAR contains the pattern 101, the output line labeled 5 from the DEC will carry a 1, with the other output lines carrying 0s. The 1 signal arriving at memory word 5 causes that register to load the contents of the MBR. Henceforth, the memory register at address 5 will contain the word 0111.

Encoders and decoders are opposites, in a sense. An encoder converts information arriving on 2^n input lines to n output lines while the decoder does the reverse. In Figures 28.2 and 28.3, we use the standard logic developed in Chapter 3.

In the case of the 2×4 encoder, it is assumed that exactly one of the four incoming lines carries a 1, with the rest being 0. The output lines carry the same information (which of d_0, d_1, d_2, or d_3 carries the 1) in binary encoded form: If $d_0 = 1$, then $b_0 = b_1 = 0$; if $d_1 = 1$, then $b_0 = 1$ and $b_1 = 0$; if $d_2 = 1$, then $b_0 = 0$

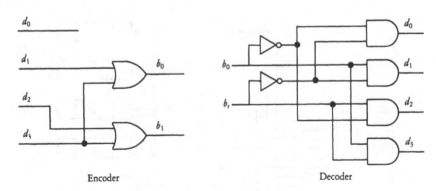

Figure 28.2   An encoder and a decoder

and $b_1 = 1$; finally if $d_3 = 1$, then $b_0 = b_1 = 1$. In all cases the binary numbers $b_1b_0 = 00,01,10,11$ encode line numbers d_0, d_1, d_2, and d_3.

The 4×2 decoder takes the same sort of binary signal as input and decodes it into one of four output lines. Each of four AND gates responds to a specific combination of input values: d_0 is 1 if $b_1b_0 = 00$; d_1 is 1 if $b_1b_0 = 01$; d_2 is 1 if $b_1b_0 = 10$; and d_3 is 1 if $b_1b_0 = 11$.

Multiplexers and demultiplexers are essentially switches which determine, for each state of a computer, which route data will travel between various components. A multiplexer has up to 2^n input lines (for some modest value of n), n control lines, and 1 output line. The signal on the control lines determines which of the 2^n input signals will be routed through the device's single output line. Shown in Figure 28.3 is a multiplexer with four input lines and the two control lines necessary to make a selection of one from four input signals.

For each combination of binary values entering the two control lines, exactly one of the four AND gates will receive two 1s. A third 0 or 1 entering that gate from its input line d_i will cause the same signal to be transmitted to the OR gate and along its output line. In fact, when s_1 and s_0 carry the binary equivalent of i, the signal on input line d_i will be transmitted through the device. Such a switching function is very useful in computers. For example, there are several possible sources, in a computer, of data to be loaded into the MBR. Each of these sources will send an input line to a multiplexer which sends its single output to the MBR. The computer's control logic will decide which source of data to use for input to the MBR at each moment. It will then send the appropriate signal along the multiplexer's control lines to connect that particular source to the MBR.

A demultiplexer or switch has a single input line and, under the direction of

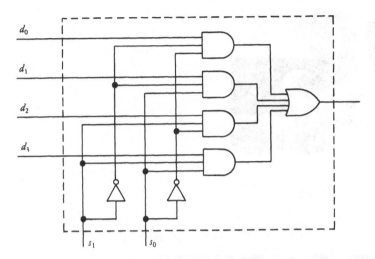

Figure 28.3 A four-input multiplexer

its control lines, selects one of its several output lines for transmission of the signal.

With encoders and decoders, multiplexers, and demultiplexers, virtually all the communication of information from one part of a computer to another can be managed.

Problems

1. Design a 3 × 8 decoder for use in the memory circuit of Figure 28.1. There will be three input bits b_0, b_1, and b_2 from the MAR.

2. Design an 8 × 3 encoder and a four-output demultiplexer.

3. Redesign the multiplexer of Figure 28.3, using only inverters, three-input NAND gates, and a single four-input NAND gate.

References

M. M. Mano. *Digital Logic and Computer Design.* Prentice-Hall, Englewood Cliffs, N.J., 1979.

Charles H. Roth, Jr. *Fundamentals of Logic Design,* 2d. ed. West, St. Paul, Minn., 1979.

29

CAT SCANNING

Cross-Sectional X-Rays

In the last two decades, medicine has been revolutionized by the appearance of computerized axial tomography (CAT), a technique for reconstructing two-dimensional cross-sectional images of patients from a multitude of one-dimensional X-ray scans. The advantages of such a facility are obvious. Rather than examine vague shadows on a traditional X-ray photograph, doctors may examine pathological features of anatomy with almost the same degree of clarity as if the patient had been cut in two!

The reconstruction problem is easy to state and understand. We do this initially in terms of a schematic cross section of a human body, and we imagine a rigid array of parallel X-ray beams and detectors which can be rotated about an axis that is somewhere in the middle of the body and perpendicular to this page (Figure 29.1).

In fact, as the apparatus rotates, a series of "snapshots" of the incoming X-ray beams X is taken by the detector array D, each image being essentially a one-dimensional strip representing the attenuation of the parallel beams by various body tissues. As in ordinary X-ray photographs, bone with its heavier elements like calcium absorbs more X-rays than does, say, lung tissue with its lighter

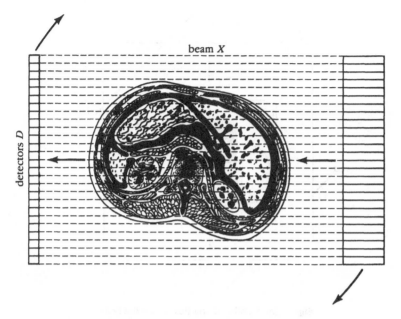

beam X

detectors D

Figure 29.1 A one-dimensional X-ray scan

composition. However, as the generator/detector array swings about the patient, the snapshots continually vary as the relative position of internal organs and tissues changes with respect to the beams: When the rays are horizontal, the spine will appear at the bottom of the image; but when the rays are vertical, the spine will show up in the middle of the image.

The detectors do not actually take a photograph of the incoming X-rays. Instead, they generate an electric signal proportional to the strength of the received ray. This signal, along with array position information, is sent to a digital computer to analyze the separate snapshots and to reconstruct an image of the distribution of materials in the body which caused the variations. Generally speaking, these materials vary from organ to organ and tissue to tissue. One can look at a reconstructed image and readily identify lungs, heart, liver, ribs, and so on.

Consider now a much simpler reconstruction problem involving a cubic solid of uniform density enclosed in a rectangular box (Figure 29.2). Three separate one-dimensional snapshots of a parallel X-ray beam are taken, one from the side, one from the top, and one from a 45° angle.

The views of the square within a box taken from A, B, and C, respectively, all look somewhat different from each other and can be represented by three functions of one variable (Figure 29.3).

193

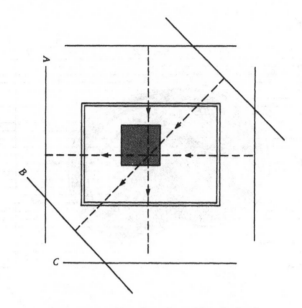

Figure 29.2 What shape lies within the box?

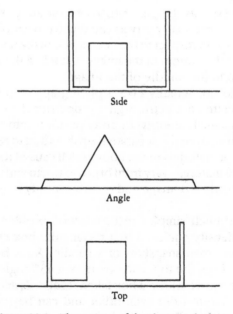

Side

Angle

Top

Figure 29.3 Three views of the shape in the box

The one-dimensional views from the top and side look rather similar with two high, sharp peaks representing the sides of the box viewed edge on and a square shape between them representing the cube. The view from an angle, however, looks rather different; the peaks have disappeared, and the cube has turned into a triangle. This latter feature is easily explained; rays traversing the thickest part of the cube travel from corner to corner while adjacent rays traverse less and less of the cube in a manner that falls off linearly with distance. Rays that do not even pass through the cube, however, still must traverse both sides of the box and do so at an angle. This explains why the box looks "thicker" in the angled view than it does in the other two views.

The simplest technique for reconstructing X-ray images is known as *back projection*. All other techniques, one of which we mention later, amount to sophisticated versions of the same thing. Back projection is easy to illustrate graphically, especially in the case of the cube-in-a-box example above. Because the number of views taken of this object was severely limited, the reconstruction will not be especially impressive.

In visual terms, we superimpose three images on each other (Figure 29.4) each image being a series of parallel bands whose density or shading represents the height of the three attenuation functions *A*, *B*, and *C*.

Although it is not at all difficult to pick out both the box and the cube within it in the image above, the figure is marred by streaks: Each back-projected density is certainly welcome wherever it contributes to the density of the object which originally produced it, but is most unwelcome elsewhere. To get rid of some of this noise, we must first divide all the densities by the number of images used in the reconstruction, in this case, 3. But to get rid of the rest, a filtering technique must be used.

If we could afford the luxury of a "continuous computer," one that could somehow deal with every single mathematical point (x, y) inside the region R of interest, then a projection could be specified as follows:

$$P(\theta, t) = \int_R f(x, y) \; \delta(x \sin \theta - y \cos \theta - t) \; dx \, dy$$

Here, $f(x,y)$ represents the tissue image to be reconstructed and $P(\theta,t)$ represents the attenuation measured for each ray at position t and angle θ. The function δ is a special filtering function applied along the ray having the (linear) equation

$$x \sin \theta - y \cos \theta = t$$

This ray makes an angle θ with the horizontal and passes within t units of the

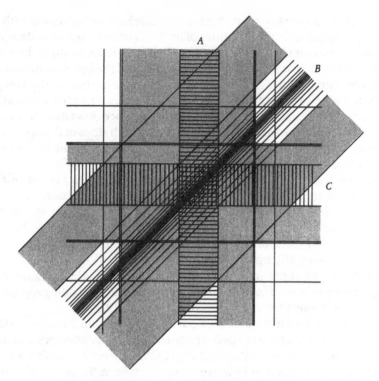

Figure 29.4 Back-projecting the three views

origin of whatever axis system is employed relative to the image. Indeed, t is merely the position in the detector array at which this ray is collected (Figure 29.5).

If θ is fixed, the function $P(\theta,t)$ therefore gives the degree of attenuation encountered for each ray (t value) in terms of the density $f(x,y)$ of all the points on it; these are simply the points which satisfy the equation $x \sin \theta - y \cos \theta - t = 0$. In this continuous case, the filtering function δ has an especially simple form: It has value 1 whenever its argument has value 0. Everywhere outside of the ray at t, δ has the value 0, so no point outside this ray may contribute to its attenuation.

If we could somehow have all possible values of $P(\theta,t)$ in hand, how could we construct $f(x,y)$ from them? In continuous terms, we would simply have to invert the foregoing integral:

$$f(x,y) = \frac{1}{2\pi} \int_0^\pi \int_{-r}^{+r} p(\theta, t)\, \delta(x \sin \theta - y \cos \theta - t)\, dt\, d\theta$$

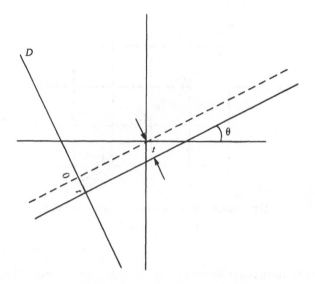

Figure 29.5 The path of an X ray

This integral instructs us to add all the contributions to the point (x,y) by the back projections at each possible angle θ from 0 to π. For a given θ, moreover, we are only interested in the t value which satisfies

$$x \sin \theta - y \cos \theta - t = 0 \qquad \text{where} - r \leq t \leq r$$

This is merely the back-projected ray which passes through (x,y).

Plainly then, if there are an infinite number of projections at our disposal, the cross-sectional X-ray density function f (and, therefore, the patient's anatomy) can be reconstructed perfectly.

Difficulties arise in the computation of image reconstructions precisely because we are forced to deal with a finite number of projections. In trying to reassemble f, we must use a discrete analog of the integral equation above:

$$f(x_i, y_j) = \frac{1}{n} \sum_{k=1}^{n} \sum_{l=1}^{m} p(\theta_k, t_l)\, \delta(x_i \sin \theta_k - y_j \cos \theta_k - t_l)$$

Here (x_i, y_j) is one of a finite set of points in the patient's body at which we are trying to reconstruct f, while θ_k and t_l are the angles and rays, each drawn from finite sets, which contribute to the reconstruction. The precise source of diffi-

197

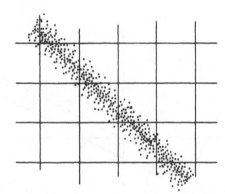

Figure 29.6 An X ray intersects pixels variously.

culty lies in the noncoincidence of any candidate set of points (x_i, y_j) and the intersections of any given finite system of rays.

In practice, the reconstructed image, like all computer-generated images, is made up of "pixels," a lattice of tiny squares within each of which the density is assumed to be uniform. For each projection angle and for each ray, even the pixels which are intersected by it have their contribution altered by the extent to which the ray intersects it. The same thing is true of the back-projected ray (Figure 29.6).

At this point we are forced to use for δ a filter function d which tells the computer precisely how much weight to give the back projection $p(\theta_k, t_l)$ at a given pixel (x_i, y_j). This filter not only must take into account the discreteness of the rays, the spacing between them, and the geometry of the pixels, but also compensate for the fact that pixels near the axis of the projections (the center of their rotation) are more heavily saturated than pixels farther from this point.

In truth, the study of tomographic reconstruction techniques is very largely the investigation of various filters, and at this point the subject becomes a bit too technical for a chapter of this length. But if such a filter function is denoted $d(x_i, y_j, \theta_k, t_l)$, then it is easy to specify a reconstruction algorithm that employs it:

CATSCAN

1. **for each** i **and** j
    $f(x_i, y_j) \leftarrow 0$
2. **for** $k \leftarrow 1$ **to** n
    for $l \leftarrow 1$ **to** m

for each i **and** j
$$f(x_i,y_j) \leftarrow f(x_i,y_j) + p(\theta_k,t_i)\ d(x_i,y_j,\theta_k,t_i)$$
3. **for each** i **and** j
$$f(x_i,y_j) \leftarrow \tfrac{1}{nm} \cdot f(x_i,y_j)$$

More sophisticated reconstruction algorithms use Fourier transforms (see Chapter 32). Such algorithms operate by first performing the discrete analog of the equation

$$P(\theta, z) = \int_{-\infty}^{+\infty} e^{-jzt} p(\theta, t)\ dt$$

and then using the inverse Fourier transform

$$f(x, y) = \frac{1}{4\pi^2} \int_{-\infty}^{+\infty} \int_{-\infty}^{+\infty} p(u, v)\, e^{j(ux+vy)}\ du\ dv$$

in its polar form, namely,

$$f(x, y) = \frac{1}{4\pi^2} \int_{0}^{\pi} d\theta \int_{-\infty}^{+\infty} P(\theta, z)\, e^{jz(x\sin\theta - y\cos\theta)} |\omega|\ dw$$

where j is the imaginary number $\sqrt{-1}$ and $|\omega|$ is a special determinant called the *Jacobian*.

Other forms of radiation diagnosis include gamma-ray tomography and nuclear magnetic resonance imaging. The computation of images for these techniques takes into account the fact that the signal originates inside the patient's body, but otherwise it bears a strong family resemblance to the reconstruction techniques outlined here.

Problems

1. Given a pixel with unit side and center at (x,y), calculate its area of intersection with a band of width w centered on the line $x \sin\theta - y \cos\theta = t$.

2. Imagine a 10×10 array of (large) pixels used to represent a patient's cross-sectional image. Consider 10 X-ray images developed at $18°$ intervals so as to comprise a complete range of scan angles to reconstruct the patient's anatomy. A simple filtering function δ may be constructed that gives, for each

pixel, (x,y) and ray (θ,t), the area of overlap between the ray and the pixel. How quickly can δ be computed "on the fly"? Would it be better to store δ as a table?

References

T. Pavladis. *Algorithms for Graphics and Image Processing.* Computer Science Press, Rockville, Md., 1982.

G. T. Herman. *Image Reconstruction from Projections.* Springer-Verlag, Berlin, 1980.

30

THE PARTITION PROBLEM

A Pseudo-fast Algorithm

O ccasionally, someone working on an NP-complete problem (see Chapter 41) imagines that she or he has discovered an exact, polynomial-time algorithm for the problem. So far, such people have turned out to be wrong! This highlights not only the need for careful analysis of new algorithms but also the fact that the property of being NP-complete can be rather subtle.

Such a problem is the *partition problem:* Given n positive integers x_1, $x_2, \ldots, x_n$, find a partition of these integers into two subsets, indexed by I and J, such that

$$\sum_{i \in I} x_i = \sum_{j \in J} x_j$$

In other words, both subsets must sum to the same amount. A simple analogy for this problem involves a set of wooden blocks of various heights (see Figure 30.1): Can we build two towers of equal height from these blocks?

The partition problem can be simplified somewhat before an algorithmic solution is attempted. For one thing, we could add the heights of all the blocks

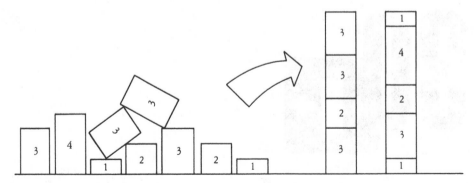

Figure 30.1 An example of the partition problem

and attempt to build a single tower whose height is one-half this sum. However, even in attempting this simpler task, we might well discover ourselves systematically trying all possible subsets of the blocks. Such a procedure would be very time-consuming. Indeed, there is no guarantee that a solution even exists, as, for example, when the sum of the heights is an odd number.

There is an algorithm which, at first sight, appears to solve the partition problem quickly. It fills in a table of truth values: the i,jth entry is a 1 (true) if there is a subset of $\{x_1, x_2, \ldots, x_i\}$ which sums to j. The algorithm scans the table, row by row. To compute the value at (i,j), it examines the values at $i-1, j)$ and at $(i-1, j-x_i)$. Obviously, if the table entry at $(i-1, j)$ is 1, then the subset of $\{x_1, x_2, \ldots, x_{i-1}\}$ which sums to j will also be a subset of $\{x_1, x_2, \ldots, x_i\}$. If, however, there is a subset of $\{x_1, x_2, \ldots, x_{i-1}\}$ which sums to $j-x_i$, then the same subset with x_i added to it sums to j.

A partial table is shown in Figure 30.2 for the blocks in Figure 30.1, taken in the order 3, 4, 3, 1, 3, 2, 3, 2, 1. These numbers sum to 22, and so we seek a subset of them which sums to 11. The first row of the table contains a single 1 since the only nonempty subset of $\{x_1\}$ is the same set and $x_1 = 3$. The next row contains three 1s corresponding to the subsets $\{x_1\}$, $\{x_2\}$, and $\{x_1, x_2\}$. These have sums 3, 4, and 7, respectively. By the time we reach the fourth row, we discover that $x_1 + x_2 + x_3 + x_4 = 11$, the required target sum.

In much the same manner as we have described this table, the algorithm generates the table row by row. To compute the value at the i,jth position, it examines the values at $(i-1, j)$ and at $(i-1, j-x_i)$. Obviously, if the table entry at $(i-1, j)$ is 1, then the subset of $\{x_1, x_2, \ldots, x_{i-1}\}$ which sums to j will also be a subset of $\{x_1, x_2, \ldots, x_i\}$ summing to j. If, however, there is a subset of $\{x_1, x_2, \ldots, x_{i-1}\}$ which sums to $j-x_i$, then the same subset, augmented by x_i, sums to j.

Here is a modified version of the algorithm in which the table entries are not 0

i \ j	1	2	3	4	5	6	7	8	9	10	11
x_1	0	0	1	0	0	0	0	0	0	0	0
x_2	0	0	1	1	0	0	1	0	0	0	0
x_3	0	0	1	1	0	1	1	0	0	1	0
x_4	1	0	1	1	1	1	1	1	0	1	1
x_5											
x_6											
x_7											
x_8											
x_9											

Figure 30.2 A tabular solution to the example

and 1 but 0 and X, where X is a subset of $\{x_1, x_2, \ldots, x_n\}$ which may differ from one entry to the next. In fact, the X at the i,jth entry will be a subset (if it exists) of $\{x_1, x_2, \ldots, x_i\}$ which sums to j.

PARTITIONFIND

1. initialize all table entries to 0
2. **for** $j \leftarrow 1$ **to** $sum/2$
       if $j = x_1$ **then** $table\,(1, j) \leftarrow \{x_1\}$
3. **for** $i \leftarrow 2$ **to** n
       for $j \leftarrow 1$ **to** $sum/2$
           if $table\,(i-1, j) \neq 0$ **then** $table\,(i, j) \leftarrow table\,(i-1, j)$
           if $x_i \leq j$ **and** $table\,(i-1, j-x_i) \neq 0$
               then $table\,(i, j) \leftarrow table\,(i-1, j-x_i) \cup \{x_i\}$
4. **if** $table\,(n, sum/2) \neq 0$
       then output $table(n, sum/2)$
       else output "No solution."

It is not difficult to prove that this algorithm is exact: It *always* finds a solution (or reports "No solution") for whatever instance of the partition problem is supplied to it. But how long does it take to do this?

Counting each **if . . . then,** assignment, and **output** statement as 1 unit of

203

time, we arrive at the following breakdown:

Line	Units of time
1	$n \times (sum/2 + 1)$
2	$sum/2 + 1$
3	n
4	$2(n - 1) \times (sum/2 + 1)$
5	1
	$3n \cdot sum/2 + 4n - sum/2 + 1$

In writing down the complexity of this algorithm, we may neglect constants and additive terms dominated by the first term. It follows that PARTITIONFIND has a complexity of

$$O(n \cdot sum)$$

At first sight, this looks as if it might be polynomial in the size of the input, so it is useful to ask, at this point, just what the size of the input is.

We have been given n integers $x_1, x_2, \ldots, x_n$. If we are sloppy and say that each integer x_i has size x_i, then we shall certainly conclude that the instance length is sum. Since $n < sum$, it will then follow that the algorithm runs in time bounded by $O(sum^2)$, which is certainly polynomial by this measure of length.

In fact, the requirement that representations of instances be "concise" rules out this way of measuring instance length: In general, it is required that an instance be represented by a scheme whose length is at most a polynomial function of a minimum-length representation. This is certainly not true of the measurement made above. In fact, the instance size of $x_1, x_2, \ldots, x_n$, when written out in binary, would be

$$s = \log_2 x_1 + \log_2 x_2 + \cdots + \log_2 x_n$$

Now if $n \cdot sum$ were bounded by a polynomial function of s, we could write

$$n \cdot sum < s^k$$

for some fixed number k and for all values of s larger than some other fixed number, m. If x_j happens to be the largest member of the set $\{x_1, x_2, \ldots, x_n\}$, then

$$s^k \le (n \log_2 x_j)^k$$

and

$$x_j \le n \cdot sum$$

It follows from the first inequality above that

$$x_j < n^k (\log_2 x_j)^k \qquad \text{for all } n \ge m$$

Since x_j and n are independent, we can choose x_j as large as we like in relation to n, even to the point where we force n to be larger than m. At this point, it is still easy to select x_j to be large enough to violate the last inequality. Therefore, $n \cdot sum$ is not bounded by any polynomial function of s, and indeed the algorithm is not a polynomial-time algorithm.

As against this sad conclusion, we might be more optimistic and ask whether the quantity "sum" might be given some kind of predefined bound, the agreement being to restrict ourselves to only those instances satisfying this bound. For example, if we require that

$$sum < s^2$$

then we admit a great many problems of "practical" interest. More precisely, there are pseudofast algorithms for scheduling problems in which the numbers are not very large (there is not much point in scheduling tasks which take billions of years).

In any event, with instances satisfying this kind of inequality, we have an algorithm which runs in time certainly no worse than $O(s^3)$, since

$$n \cdot sum < n \cdot s^2 < s^3$$

Previously we have used the term *pseudofast* in place of the more standard term *pseudopolynomial time*. An algorithm runs in *pseudopolynomial time* if its complexity on each instance I is bounded above by a polynomial in both length I and max I, where max I is just the size of the largest number in I.

The partition-finding algorithm is a pseudopolynomial-time algorithm for the partition problem since

$$n \cdot sum < n \cdot (n \cdot x)$$
$$< n^2 \cdot x$$
$$< s^2 \cdot x$$

205

where $x = \max \{x_1, x_2, \ldots, x_n\}$ and s is the length function defined above.

Finally, we note a more general problem than the partition problem. The partition-finding algorithm also solves the "subset sum problem" in pseudopolynomial time. Given the integers $x_1, x_2, \ldots, x_m$ and another integer, B, find a subset of the m integers that sums to B. This problem will be discussed in Chapter 37 on public-key encryption systems.

Problems

1. Suppose the n integers are given as 1, 2, 4, 8, $\ldots$, 2^{n-1}. For what sums B does the subset-sum problem have a solution, and how quickly can one find it?

2. If there are n integers, all the same size, it is certainly easy to solve the partition problem very quickly. Does there exist a polynomial-time algorithm when the integers have just two possible sizes — no matter how large the sum is?

References

M. R. Garey and D. S. Johnson. *Computers and Intractability: A Guide to the Theory of NP-Completeness.* Freeman, San Francisco, 1979.

David Harel, *Algorithmics: The Spirit of Computing,* Addison-Wesley, Wokingham, England, 1987.

31

TURING MACHINES

The Simplest Computers

uring machines are the simplest and most widely used theoretical models of computing. Far too slow and unwieldy ever to be embodied in a real device, these conceptual machines nevertheless seem to capture everything we mean by the term *computation*. Not only do Turing machines occupy the top level of the Chomsky hierarchy (see Chapter 7), but also they seem capable of computing any function which is computable by any other conceptual scheme (see Chapter 66). Turing machines, moreover, are simpler than such schemes—from general recursive functions to random access machines (see Chapters 17 and 48).

As with any computational device or model, we must be careful to distinguish "hardware" from "software." In the case of Turing machines, we might visualize the kind of physical setup shown in Figure 31.1.

An infinite tape composed of discrete cells is examined, one cell at a time, by a read/write head which communicates with a control mechanism inside a box. Under the control of this mechanism, the Turing machine is able to read the symbol in the current tape cell or to replace it by another. The Turing machine is also able to move the tape, one cell at a time, either to the left or to the right.

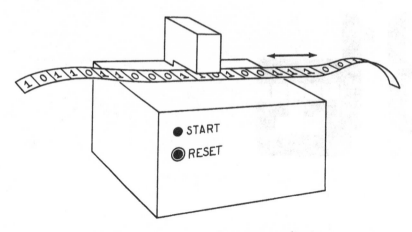

Figure 31.1 A Turing machine conceptualization

Like the other machines in the Chomsky hierarchy, Turing machines may be treated as language acceptors. The set of words which cause a Turing machine ultimately to halt is considered to be its *language*. However, as indicated above, Turing machines are much more than this. We consider the function defined by a Turing machine M as follows:

Let x be a string over M's tape alphabet Σ. Place the read/write head over the leftmost symbol of x, start M in its initial state, and allow it to compute without interference. If M eventually halts, then the string y which is found on the tape at that time will be considered as M's output corresponding to the input x. Since M does not necessarily halt for every input string x, M computes a partial function

$$f\colon \Sigma^* \to \Sigma^*$$

where Σ^* denotes the set of all strings over Σ. Under this scheme, all sorts of possibilities for Σ, the tape alphabet, may be envisaged. For example, Σ may be used as a basis for the representation of integers, rational numbers, alphabetic words, or virtually any other class of objects suitable for computation.

Strictly speaking, a Turing machine is essentially the same thing as its *program*. Formally, such a program M is a set of quintuples of the form

$$(q, s, q', s', d)$$

where q is the current state, s is the symbol currently under the read/write head, q' is the state which M is next to enter, s' is the symbol to be written in place of s, and d is the direction in which the read/write head is to move relative to the

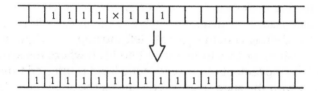

Figure 31.2 Multiplying two unary numbers

tape. The states come from a finite set Q, the symbols from an alphabet Σ, and the read/write motions from $D = \{\text{left, right, stop}\}$.

A Turing machine program, given a particular input string, will always start in a specially designated *initial state* $q_0 \in Q$ and may be caused to halt in a number of ways. The method we use to halt a program involves the use of *halting states*: When M enters a halting state q', all computational activity ceases. Naturally, q' never appears as the initial symbol in any of the quintuples of M's program.

A Turing machine program P shown below is for multiplying two numbers in unary form. Under this representation, a number m is represented by m consecutive 1s on the tape. Initially, P's tape will contain the two numbers to be multiplied separated by an X symbol. Finally, P's tape will contain the product of m and n in unary form. The initial and final tapes are shown in Figure 31.2.

As in other conceptual machines, there are several ways of representing a Turing machine program, the most direct being simply a list of P's quintuples:

MULTIPLICATION PROGRAM

$q_0, 1, q_1, 1, L$ $q_6, 1, q_6, 1, L$

$q_1, b, q_2, *, R$ $q_7, *, q_7, *, R$

q_2, b, q_3, b, L $q_7, 1, q_7, 1, R$

$q_2, *, q_2, *, R$ q_7, X, q_5, X, L

$q_2, 1, q_2, 1, R$ q_7, A, q_7, A, R

q_2, X, q_2, X, R q_8, b, q_3, b, L

q_2, A, q_2, A, R $q_8, 1, q_8, 1, R$

$q_3, 1, q_4, b, L$ q_8, X, q_8, X, R

q_3, X, q_4, X, L $q_8, A, q_8, 1, R$

$q_4, 1, q_4, 1, L$ $q_9, *, q_{10}, b, S$

q_4, X, q_5, X, L $q_9, 1, q_9, 1, L$

$q_5, *, q_8, *, R$ $q_{10}, b, \textcircled{$q_{11}$}, b, S$

$q_5, 1, q_6, A, L$ $q_{10}, 1, q_{10}, b, R$

q_5, A, q_5, A, L q_{10}, X, q_{10}, b, R

$q_6, b, q_7, 1, R$ q_{10}, A, q_{10}, b, R

$q_6, *, q_6, *, L$

The most readily understandable representation of a Turing machine program is the state-transition diagram. Such a diagram can often be simplified even further by labeling certain states as "left-moving" or "right-moving" and omitting all transition arrows from a state into itself where the symbol read is rewritten and the movement of the read/write head is inferred from the direction associated with the state. Figure 31.3 shows the state-transition diagram for P.

The first action of P, on being confronted with two unary numbers to multiply, is to move one cell to the left from the leftmost 1 and to write an asterisk there, entering state 2. Program P remains in state 2, continually moving to the right until the first blank is arrived at. From state 3 P changes the 1 beside this to a blank (state 3 $\rightarrow$ state 4) and moves left until it encounters the X sign. Now in state 5, it enters a three-state loop in which each 1 in the left-hand unary number is changed to an A and a corresponding 1 is written in the first available cell to the left of the asterisk. In this way, P simply writes a string of m ones. When P finally encounters the asterisk, the string is finished and P moves to state 8, where it changes all the A's back to 1 and then continues moving to the right until a blank is encountered (state 8 $\rightarrow$ state 3). Here, the cycle begins again, the next 1 in the string of n ones is blanked, and P reenters the three-state loop in order to write another string of m ones adjacent to the string it had just written. In this manner, n adjacent strings of m ones are written, completing the formation of the product $m \times n$ in unary form. In the final stage of its computation, P erases the asterisk (state 9 $\rightarrow$ state 10) and moves to the right, erasing all symbols (denoted by $-$) until it reaches the first blank. Then it halts.

Turing machines were first described by Alan Turing whose role as one of the founders of computer science this book celebrates. Turing conceived of his machine in the mid-1930s in an effort to write down in the simplest possible terms what it means for a computation to proceed in discrete steps. Turing suspected that his machine was the most powerful conceivable theoretical

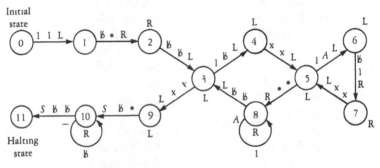

Figure 31.3 The unary multiplication program

model when he noticed that it was equivalent to a number of other conceptual computing schemes. Another factor which may have given him confidence in this thesis was the robustness of Turing machines. Their alphabets, state sets, and tapes can be changed radically without affecting the power of the class of machines so obtained.

For example, the alphabet of a Turing machine may contain as few as two symbols and still compute anything which machines with larger alphabets can compute. The number of states can also be limited to just two without altering their power! Naturally, there is a trade-off between the number of states and the size of alphabet: Given an arbitrary Turing machine program P employing m states and n symbols, a two-symbol program which mimics P must, in general, have far more than m states. Conversely, the two-state program which mimics P must have far more than n symbols.

One of the most robust features of Turing machines is the variation in the kind and number of tapes they may have without their computational powers being affected. Turing machines may have any number of tapes, or they may get along perfectly well on a single, semi-infinite tape. This is an infinite tape which has been cut in half, so to speak. If the right-hand half is used, then a Turing machine with this sort of tape is not permitted to move beyond the leftmost cell.

A *multitape Turing machine program* is not a set of quintuples, but a set of $(2 + 3n)$-tuples, where n is the number of tapes being used. The "instructions" of such a program have the following general form:

$$q, s_1, s_2, \ldots, s_n, q', s_1', s_2', \ldots, s_n', d_1, d_2, \ldots, d_n$$

where q is the current state and $s_1, s_2, \ldots, s_n$ are the symbols currently being read (by n read/write heads) on the n tapes. Under these conditions, this $(2 + 3n)$-tuple tells us that the machine must next enter state q', write the symbols $s_1', s_2', \ldots, s_n'$ on their respective tapes, and move the corresponding read/write heads in directions $d_1, d_2, \ldots, d_n$, where, of course, each d_i may be L, R, or S as in the one-tape machine described earlier.

Although they cannot compute anything which a one-tape machine cannot compute, multitape machines can often compute much more efficiently. For example, given the task of forming the product of two unary numbers, there is a three-tape machine program which does this in approximately $n \times m$ steps. The program presented earlier requires in the order of $n^2 \times m^2$ steps!

We close this chapter with a demonstration that a one-tape Turing machine can do anything that an n-tape machine can do. More precisely, we show that given any n-tape Turing machine program M, there is an $(n-1)$-tape Turing machine program P which does precisely the same thing. The basic construction can then be reapplied to produce an equivalent $(n-2)$-tape machine and so on. Ultimately a one-tape machine that is equivalent to M results.

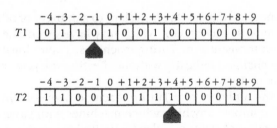

Figure 31.4 The two tapes of *M*

For the sake of simplicity, suppose that *M* is a two-tape Turing machine. Denoting the two tapes by *T*1 and *T*2, select an arbitrary cell in each as cell 0 and number all cells to the right and left of these cells with positive and negative numbers, respectively (Figure 31.4).

On a new tape, number the cells in pairs 0,0, then +1,+1, and so on, placing the symbol from the *i*th cell of *T*1 in the first cell of the *i*th pair and placing the symbol from the *i*th cell of *T*2 in the second cell of the *i*th pair (Figure 31.5). It is convenient to assume that *M*'s tape alphabet consists only of the binary numbers 0 and 1. The positions of the two read/write heads are indicated by alphabetic symbols *A* and *B*. These symbols serve only to indicate 0 and 1, respectively. Which head (1 or 2) they represent is kept track of by the Turing machine *M'* that we will presently construct.

The general idea behind the construction of *M'* is to have its single read/write head mimic the read/write heads of *M* by alternating between the two markers. Each marker is therefore moved two cells at a time since cells that are adjacent on one of the two original tapes of *M* are now two cells apart on the single tape of *M'*. The new Turing machine *M'* will be specified in terms of a state-transition diagram that has two symmetric halves. In one half, the one in Figure 31.6, the read/write marker for tape 1 is assumed to lie to the left of the read/write marker for tape 2. If, in the course of its simulation of *M*, the two markers of *M'* should happen to cross, a transition takes control from the first half to the second; here, the tape 1 marker is assumed to lie on the right of the tape 2 marker.

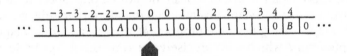

Figure 31.5 Merging the two tapes into one

The first step in the operation of M' is to decide which 8-tuple of M must be "applied" next:

$$q, s_1, s_2, q', s_1', s_2', d_1, d_2$$

Given that M is supposed to be in state q, we assign the same state to M' and enter a portion of the state-transition diagram that looks like a decision tree. The read/write head of M' is currently scanning the left-hand marker (Figure 31.6).

In this and subsequent diagrams, some simplifying conventions are introduced: If no symbol appears in the middle of a transition arrow, the symbol at the beginning is merely rewritten. A state labeled L or R is left-moving or right-moving, respectively. In other words, M' keeps moving its read/write head in the indicated direction, remaining in that state, until a symbol appearing on one of its outgoing transition arrows is encountered. Arrows with two symbols at their beginning actually indicated two arrows, one per symbol.

The machine M' first branches to one of the two right-moving states depending on whether the left-hand marker is an A (0) or B (1). It continues moving to the right in either state until it encounters the right-hand marker. Here, a similar decision is made, and the read/write head now moves back to the left-hand marker prior to a transition out of the diagram, so to speak.

At this stage there is enough information to specify which transition

$$q, s_1, s_2, q', s_1', s_2', d_1, d_2$$

is to be simulated by M. The symbols s_1 and s_2 are both known. They are represented by the values (A or B) of the two markers. The diagram that handles the writing of s_1', s_2', as well as the marker movements indicated by d_1 and d_2, has a very simple structure (Figure 31.7).

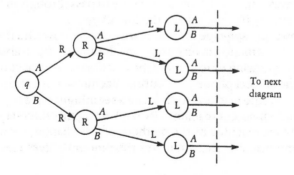

Figure 31.6   Reading the two markers

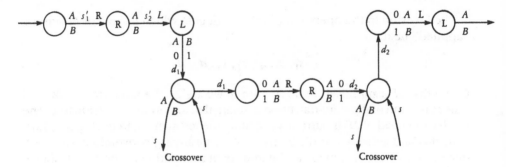

Figure 31.7 Writing and moving the markers

At the left-hand marker, M' replaces A or B by an alphabetic version of the symbol s_1', then moves to the right until the right-hand marker is encountered. This marker is replaced by an alphabetic version of s_2', and the read/write head of M' moves back to the first marker, replacing it by the corresponding numeric symbol. Then M' moves one step in direction d_1 and checks to see if the current cell holds an A or a B. If so, the markers are about to cross: An appropriate transition thus carries M' into the other half of its diagram, the one for which marker 1 lies to the right of marker 2. Naturally, there is a corresponding transition to this state from the corresponding state of the other half diagram. In any event, M' makes another move in direction d_1 and replaces the numeric symbol in this cell by its alphabetic counterpart. The left-hand marker has been moved two cells in direction d_1.

Next M' carries out the corresponding operation on the right-hand marker and finally moves back to the first marker. As soon as this marker is encountered, M' undergoes a transition that takes it out of the diagram above. Which state does M' enter next? It enters q', a state "borrowed" from M in precisely the manner that q was borrowed. Now M' is ready to pass through another decision tree of states exactly like the one in the first diagram.

Besides showing that n-tape machines are no more powerful than single-tape machines, this construction is very useful in simplifying much abstract theory of computation by allowing us to confine the discussion of what computers can or cannot do to the subject of one-tape Turing machines. For example, in Chapter 66, we use a three-tape machine to simulate a seemingly more powerful abstract device, the random-access machine. By converting this three-tape Turing machine to a one-tape machine by the methods of this chapter, we would obtain a one-tape Turing machine that simulates the random-access machine in question.

Problems

1. Show that a Turing machine with a semi-infinite tape can carry out any computation of which the standard Turing machine is capable.

2. Supply the details for the conversion of an n-tape Turing machine to an equivalent $(n - 1)$-tape machine by adapting the 2-to-1 conversion to move complex tuples of the n-tape machine.

3. Write a three-tape Turing machine program for multiplying two binary numbers.

References

M. L. Minsky. *Computation: Finite and Infinite Machines.* Prentice-Hall, Englewood Cliffs, N.J., 1967.

H. R. Lewis and C. H. Papadimitriou. *Elements of the Theory of Computation.* Prentice-Hall, Englewood Cliffs, N.J., 1981.

THE FAST FOURIER TRANSFORM

Redistributing Images

When one is confronted with a computational problem involving a certain mathematical object X, it is sometimes possible to transform X to another object and to solve a (related) computational problem on the newly transformed object. For example, when X is a real number, one may take its logarithm before finding its product with another number Y; instead of multiplying X and Y, one adds their logarithms, taking the antilogarithm of the result:

$$X \cdot Y \rightarrow \log X + \log Y \rightarrow \text{antilog } (\log X + \log Y) = X \cdot Y$$

When X is a real-valued function, it is useful in certain problems to take the Fourier transform of X before proceeding. Fourier transforms are used frequently in the processing of images by computer; in Figure 32.1, for example,

Figure 32.1 An image and its Fourier transform

the Fourier transform of the white square shown at the left is the pattern shown at the right. In essence, the Fourier transform of an image redistributes it throughout the picture plane. One use of this property is that distortions or errors in the transformed image may not be visible when the latter picture is transformed to the original image.

The Fourier transform of a continuous real-valued function f is the integral

$$F(u) = \int_{-\infty}^{+\infty} f(x) e^{-iux} \, dx$$

In computing a Fourier transform, we must use the discrete version

$$F(k) = \sum_{j=0}^{n-1} f(j) \omega^{kj}$$

where $f(j)$ is a vector $(f(0), f(2), \ldots, f(n-1))$ and ω is the nth root of unity, namely,

$$\omega = e^{2\pi i/n},$$

where i is the imaginary number $\sqrt{-1}$. A more compact way of writing the discrete Fourier transform is

where f^T is the transpose of vector f and A is the $n \times n$ matrix the kjth element of which is ω^{kj}. Matrix A turns out to be nonsingular, and the kjth element of its inverse A^{-1} is ω^{-kj}/n. We can, therefore, define the inverse Fourier transform of

217

a vector g by

$$F^{-1} = A^{-1} g^T$$

In fact, if we take the Fourier transform of f and follow this by the inverse transform, we obviously get f again:

$$f^T = A^{-1} A f^T$$

Not mentioned until now is the fact that ω is a complex number and therefore the Fourier transform of f has two parts, a real part and an imaginary part. In many applications, only the real part is used.

A straightforward program for computing the Fourier transform of a function (i.e., vector) f would probably proceed by multiplying each row of A by the column f^T, using on the order of n multiplications each time. This leads to a total of n^2 such operations to complete the transform. When n is rather large (such as it is bound to be in many applications), n^2 operations become something of a burden to perform, and even in the early 1960s faster methods were sought.

In 1965, J. M. Cooley and J. W. Tukey discovered a method of computing the discrete Fourier transform which required on the order of only $n \log n$ operations. Their method uses a divide-and-conquer strategy, so common in optimally fast algorithms.

An attractive device for understanding how Cooley and Tukey's fast Fourier transform works is to view the problem of computing

$$F(k) = \sum_{j=0}^{n-1} f(j) \omega^{kj} \quad \text{at } k = 0, 1, \ldots, n-1$$

as a problem of evaluating the polynomial

$$P(x) = \sum_{j=0}^{n-1} f(j) x^j$$

at $x = \omega^0, \omega^1, \ldots, \omega^{n-1}$. But to compute $P(\omega^k)$, it is only necessary to divide $P(x)$ by $x - \omega^k$. The theory of polynomials tells us that the remainder of this division will be $P(\omega^k)$. It follows that the problem of computing P at these n values is just a matter of finding n such remainders.

The divide-and-conquer strategy enters the computation at precisely this point: Assuming for the time being that n is a power of 2, say 2^m, form the quantities $x - \omega^k$ into two products containing $n/2$ factors each. Divide $P(x)$ by each of these products:

$$\frac{P(x)}{(x - \omega^0)(x - \omega^1)(x - \omega^2)(x - \omega^3)} \qquad \frac{P(x)}{(x - \omega^4)(x - \omega^5)(x - \omega^6)(x - \omega^7)}$$

The two divisions will result in two quotients, $Q_1(x)$ and $Q_2(x)$. These may now be divided by products containing $n/4$ factors each. These factors are different from those which produced the respective quotients.

$$\frac{Q_1(x)}{(x - \omega^4)(x - \omega^5)} \qquad \frac{Q_1(x)}{(x - \omega^6)(x - \omega^7)}$$

$$\frac{Q_2(x)}{(x - \omega^0)(x - \omega^1)} \qquad \frac{Q_2(x)}{(x - \omega^2)(x - \omega^3)}$$

For example, since $x - \omega^0$ was part of the division producing $Q_1(x)$, it does not participate in any further division involving this quotient. Once again we divide the resulting quotients $Q_{11}(x)$, $Q_{12}(x)$, $Q_{21}(x)$, and $Q_{22}(x)$ by products, this time by ones which contain $n/8$ factors each:

$$\frac{Q_{11}(x)}{x - \omega^6} \qquad \frac{Q_{11}(x)}{x - \omega^7} \qquad \frac{Q_{12}(x)}{x - \omega^4} \qquad \frac{Q_{12}(x)}{x - \omega^5}$$

$$\frac{Q_{21}(x)}{x - \omega^2} \qquad \frac{Q_{21}(x)}{x - \omega^3} \qquad \frac{Q_{22}(x)}{x - \omega^0} \qquad \frac{Q_{22}(x)}{x - \omega^1}$$

In this particular example, where $n = 2^3$, there were only three iterations of this procedure. It is not hard to see, however, that by the time quotients are being divided by trivial products of the form $x - \omega^k$, there have been m iterations and the remainder from each of these final n divisions is in each case the same as the remainder from the division

$$\frac{P(x)}{x - \omega^k} \qquad k = 0, 1, \ldots, n - 1$$

If $P(x)$ had been divided singly by these n factors, there would have been on the order of n operations carried out for each division, resulting in the order of n^2 operations overall. However, it turns out that each division at the top level can be performed with $n/2$ operations, each division at the next level with $n/4$, the next with $n/8$, and so on. Thus at each iteration there are on the order of n operations to be performed for a total time complexity of

$$m \cdot n = n \log n$$

Before the fast Fourier transform can be presented algorithmically, there are some loose ends to clean up. The first involves the observation that a polynomial $Q(x)$ having degree $2^{k+1} - 1$ can be divided by a polynomial

$$(x - \omega^i) \cdots (x - \omega^j)$$

of degree 2^k in just 2^k elementary steps. Due to the special nature of the powers of ω, certain products of the factors $x - \omega^j$ can be written very simply. Specifically,

$$(x - \omega^{r(j)})(x - \omega^{r(j+1)}) \cdots (x - \omega^{r(j+2^k-1)}) = x^{2^k} - \omega^{r(j/2^k)}$$

where r is a function that maps each integer into the one having the same binary digits in its expansion but in reverse order. The division

$$\frac{Q(x)}{x^{2^k} - \omega^{r(j/2^k)}}$$

can obviously be performed within a constant multiple of 2^k steps.

Earlier it was assumed that $n = 2^k$, but what if n is not a power of 2? In many applications, especially those involving picture processing, n is arranged to be a power of 2. In other instances, however, the function f is "padded" with additional values, say 0s, until it becomes a vector of the appropriate length.

An algorithm for the fast Fourier transform follows:

FOURIER

1. for $j \leftarrow 0$ **to** 2^{k-1}
  $Q(j) \leftarrow Q_j$
2. for $l \leftarrow 0$ **to** $k - 1$
  1. for $j \leftarrow 0$ **to** 2^{k-1}
    $S(j) \leftarrow Q(j)$
  2. for $j \leftarrow 0$ **to** 2^{k-1}
    compute $(d_0 d_1 \cdots d_{k-1})$, the binary representation at j
    $Q(j) \leftarrow S(d_0 \cdots d_{l-1} 0 d_{l+1} \cdots d_{k-1})$
      $+ \omega^{d_l d_{l-1} d_0 0 \cdots 0} \cdot S(d_0 \cdots d_{l-1} 1 d_{l+1} \cdots d_{k-1})$
3. for $j \leftarrow 0$ **to** $2^k - 1$
  $b(j) \leftarrow Q(r(j))$

In this algorithm, Q is an array that stores the coefficients of the quotient polynomials described earlier. In statement 2, the levels of iteration from the

0th to the $(k-1)$st are set up. Obviously, there will be $k = \log n$ of these. At line 2.1 a temporary array S is initialized. At line 2.2 the inner loop of n iterations is set up, and within this loop the divisions are performed. In fact these divisions are performed so efficiently (as a result of mathematical simplification) that the jth coefficient at the lth iteration is essentially a linear combination of two entries in array S, with both indices computed very simply by a simple alteration of the binary representation of j.

In Fourier transforms of digitized $n \times n$ pictures, one uses the formula

$$F(k, l) = \frac{1}{n} \sum_{i=0}^{n-1} \sum_{j=0}^{n-1} f(i,j) \omega^{(ik+jl)/n}$$

where $f(i, j)$ is a picture intensity value at the i,jth location within the image grid. This is obtained from our earlier formula for $F(k)$ by the coordinatization of the indices k; these are replaced by the double indices (k, l), and the coefficients and powers of the resulting formula are adjusted accordingly. The algorithm for performing the fast Fourier transform for this two-dimensional formula is very similar to the algorithm given above.

A specific and easily described application of the fast Fourier transform involves the multiplication of two polynomials

$$a(x) = \sum_{i=0}^{n-1} a_i x^i \quad \text{and} \quad b(x) = \sum_{j=0}^{n-1} b_j x^j$$

Briefly,

$$a(x) \cdot b(x) = F^{-1}(F(a) \cdot F(b))$$

for the Fourier transform F of polynomial a (represented by its coefficient vector) is nothing more than the array of a's values at the n roots of unity ω^0, $\omega^1, \ldots, \omega^{n-1}$. The products of these values, say $a(\omega^j) \cdot b(\omega^j)$, gives the values of $a(x) \cdot b(x)$ at the same roots of unity. It follows that the inverse Fourier transform F^{-1} translates the roots-of-unity representation of the product polynomial back to its representation as a coefficient vector.

Other uses of the fast Fourier transform occur in the digital processing of engineering, medical, geological, and virtually all kinds of scientific data in which one-, two-, or even higher-dimensional data can be represented by arrays of observational values.

Problems

1. Compute the Fourier transform of the following one-dimensional "picture":

$$0\ 1\ 0\ 1\ 0\ 1$$

2. Use a hand-generated Fourier transform and its inverse to generate the product of the polynomials

$$2x^2 - 1.5x + 3 \quad \text{and} \quad x^2 + 3x - 0.35$$

Warning: the exercise is lengthy and not recommended as a shortcut. Readers with access to an appropriate FFT program may wish to use that, instead.

References

A. Aho, J. Hopcroft, and J. D. Ullman. *The Design and Analysis of Computer Algorithms.* Addison-Wesley, Reading, Mass., 1974.

William H. Press, Brian P. Flannery, Saul A. Tenkolsky, and William T. Vetterling. *Numerical Recipes: The Art of Scientific Computing.* Cambridge University Press, Cambridge, 1986.

33

ANALOG COMPUTATION

Spaghetti Computers

In computer books of the 1950s and 1960s, it was not uncommon to find two sections, one on digital computers and the other dealing with analog machines. As time went on, the analog section got smaller and smaller, being reduced eventually to a few footnotes and then disappearing altogether. The lack of interest in analog computers is surely a reflection of their displacement by the ultimately faster and more flexible digital computer, but at one time, mainly in the 1930s and 1940s, most of the real number crunching, for example, the computation of ballistics tables, was carried out by mechanical analog computers called *differential analyzers*.

There are other kinds of analog computers that are usually described in a very informal way and never seriously proposed in hardware terms. These are conceptual models of analog computation which, like all analog devices, are designed to solve one specific class of problems. Such models have the very dramatic quality of solving problems in one swoop, as it were, and are usually embodied in the homliest terms as a bunch of strings, nails in a board, or what have you. Most of the examples that follow are part of computer science folklore, passed around by word of mouth and rarely described in the literature.

Our first example involves a handful of uncooked spaghetti, individual rods having been trimmed to various lengths which reflect the sizes of numbers one wishes to sort. For example, if the numbers are not too large, say, 7, 15, 4, 18, 11, 10, 21, 16, 9, 12, 20, 13, then the spaghetti rods could be trimmed to these lengths, measured in centimeters. One picks them up in a bundle with one hand and, holding them fairly loosely, slams them end-down onto a tabletop (Figure 33.1).

At one stroke the problem of sorting these numbers has been solved; all that needs to be done now is to pick the rods out of the bundles one at a time. The longest rod is easiest to spot, grab, and remove. Merely repeat this operation, recording the lengths, until the bundle is exhausted. The numbers have now been sorted into decreasing order!

The next example involves finding the shortest path in a graph. Formally, a *graph* is defined as a finite set of elements called *vertices* and a set of (unordered) pairs of these vertices called *edges*. When the edges of a graph are assigned specific numbers (interpreted as lengths) and two vertices, u and v, are specified, then we may ask for the shortest path in the graph from u to v.

Once again, we may solve this shortest-path problem by preparing a physical analog of the graph in question. Cut a long string into lengths which reflect the numbers assigned to edges, allowing a little extra for knotting the ends of the strings together. When the strings have thus been tied into a configuration

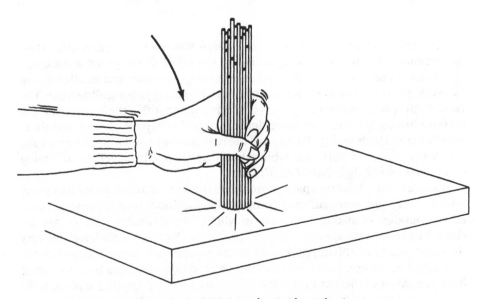

Figure 33.1 Sorting numbers with spaghetti

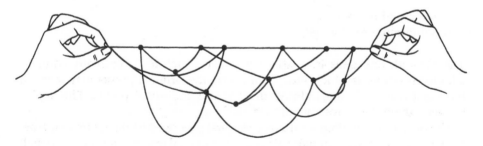

Figure 33.2 Finding the shortest path in a (string) graph

identical to the graph, take the "vertices" (knots) u and v in separate hands and pull them apart until the network of strings resists any further separation (Figure 33.2).

The shortest path (or all shortest paths, if there is more than one) will stand out clearly as a sequence of taut strings between the two hands. If knots have been labeled with the names of the vertices which they represent, then the shortest path can easily be identified in the original graph and the problem is solved.

Sorting numbers is a problem that arises in all kinds of commercial computing situations. Finding the shortest path in a graph has a few applications in various network environments involving interconnected computers or trans-shipment problems. The next problem has even fewer applications but is interesting, nevertheless, to researchers in geometric computing. Given a finite set of points in the plane, find its convex hull, that is, the convex region of minimum area which contains all the points. Clearly, a number of these points not only will lie on the boundary of the convex hull, but also will serve to define it as a cyclic sequence of straight-line segments joining the boundary points in some order. This suggests the following analog solution of the convex-hull problem.

Select a large sheet of plywood to represent the plane, and drive in nails to represent the points. Again, care must be taken in the construction so that the placement of the nails accurately reflects the positions of points in a given instance of the convex-hull problem. When the nails have all been driven, select a reasonably flexible rubber band, stretch it around the configuration of nails, and allow it to snap into place (Figure 33.3).

The next analog "machine" is somewhat spectacular in that it not only produces solutions to a problem more difficult than any of the foregoing, but also can actually be built from simple materials. The problem in question is the Steiner minimum-tree problem: Given a finite collection of points in the plane, find a network which

Connects all the points
Has minimum total length

The network one finds invariably has a tree structure. It normally contains additional points at which lines connecting the original points may meet. It turns out that three lines always meet at the additional points. The angles between them, moreover, are always 120°.

The problem of finding such a tree was first proposed by the German mathematician Jacob Steiner. Initially of purely theoretical interest, it now has a small number of applications, including the calculation of long-distance telephone charges. At least, it could be more easily applied in this situation if only someone could find a fast, exact computer program to solve this problem. No one has.

To solve the Steiner minimum-tree problem, construct a special holder consisting of a wooden handle piece and two sheet of rigid, clear Plexiglas. Given a set of points in the plane which one wishes to connect by means of a Steiner minimum tree, insert rubber-tipped pegs between the sheets to represent the points, dip the apparatus in a soap solution, and then lay it flat on a table—or on an overhead projector to obtain interesting effects on your wall (Figure 33.4).

The soap solution connects the pegs in a tree which, as the film glides into its equilibrium position, attains minimum total length. This is because soap films tend to develop a surface of minimum total area in accordance with the natural principle of minimum energy.

Unfortunately, the minima attained by soap films in such an apparatus are not

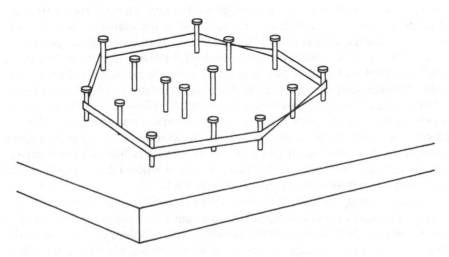

Figure 33.3 A rubber band traces the convex hull.

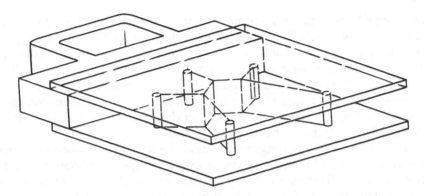

Figure 33.4 A gadget to find Steiner minimum trees

always the true, overall minima, but merely what mathematicians call a *local minimum*. Depending on the angle of dip and the particular peg configuration used, the film in equilibrium may develop a true minimum or only a local minimum.

The final analog device described in detail involves a set of parallel mirrors and two lasers (Figure 33.5). One of the mirrors, $M1$, is fixed in position and is lined with a photoelectric detection strip which responds locally to the coincidence of the two (reflected) laser beams on a given spot. At the left end of $M1$ is a detector D which works in the same way. The other mirror, $M2$, is adjusted so that a fixed reference laser R reflects, say, $n-1$ times from $M1$ and is absorbed at D. A second, movable laser P is rotated so that

Its beam always crosses the R beam at the right-hand edge of $M1$.
It sweeps out an angle from $0°$ (with the horizontal) to the same angle made by R.

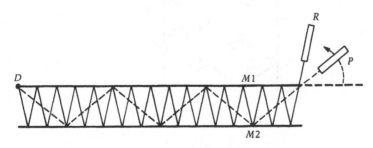

Figure 33.5 A laser and mirrors detects prime numbers.

When the probe beam P finally comes to rest, it coincides with the R beam, and in the process of its rotation, it may yield a reflection on $M1$ at the same point where the R beam reflects and, at the same time, be absorbed at D. In such a case, the number n is not prime but a composite, and the P angle at which the coincidence occurs is easily translated to an integer m which is a factor of n.

If, however, in the process of sweeping out its angle, the P beam creates no coincident reflection/absorption of the sort described, then the number n is prime and has no factors other than 1 and itself.

Of course, one dare not suggest that any of the devices described so far have any practical value. Although each one could be built, it would only succeed in solving rather small instances of the problem it was designed for. If a great many numbers are to be sorted, the spaghetti method becomes cumbersome. If one wants to find the shortest path in a very large graph, the weight of the strings would become so great that the shortest path could not even be pulled taut. A very large number of nails driven into a sheet of plywood is likely to have a convex hull in which it would be unclear whether some of the nails were really touching the elastic. The soap solution to the Steiner minimum-tree problem does not always work, even in principle, and inaccuracies in the mirrors and reflective absorption of the laser beams would make the photomechanical prime finder useless for primes much larger than 100, if not smaller.

As against these objections, however, it is nevertheless interesting to treat the devices described here in an idealized way and to ask which computations are "in principle" achievable. With the exception of the Steiner minimum-tree problem, all are—and much else besides. For example, if one aims an idealized pulsed laser into a square enclosure lined with perfect mirrors, it does not take a lot of mental work to discover that a pulse will ultimately escape through the pinhole by which it first entered the box only if the angle of the laser has rational slope (Figure 33.6). Here, then, is an analog scheme for distinguishing a

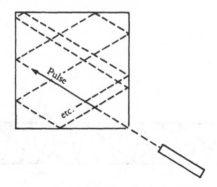

Figure 33.6 The light-in-a-box problem

rational number from an irrational one, something that cannot be contemplated by a digital computer even in principle!

We end this topic with an interesting observation: The only problem treated here for which the analog solution failed in principle (in some cases) is the Steiner minimum-tree problem—which just happens to be NP-complete (see Chapter 41). The latter property means, in effect, that no efficient, exact algorithmic solution is known for this problem or is ever likely to be known. Is it possible that none of the really difficult discrete computational problems have any analog solutions either? Whatever the answer, it would seem wise to remind ourselves that digital computers may be only a step toward new, even more powerful computational devices embodying new principles (whether analog, digital, or something quite different) that we have not yet dreamed of.

Problems

1. It is possible to make a very nice pentagon by "tying a knot" in a narrow strip of paper with parallel sides. Try this. Then prove that with an idealized strip of paper subsequently knotted and flattened, the result is indeed a true pentagon.

2. Describe an analog method to find the center of mass of an idealized solid of constant thickness.

3. Prove that the laser pulse eventually leaves the mirror box if and only if the angle of the laser has rational slope with respect to the horizontal.

References

H. H. Goldstine. *The Computer: From Pascal to Von Neumann.* Princeton University Press, Princeton, N.J., 1973.

C. Isenberg. *The Science of Soap Films and Soap Bubbles.* Tieto, Avon, England, 1978.

34

SATISFIABILITY

A Central Problem

The satisfiability problem in boolean algebra appears, at first sight, to be rather simple: Given an expression like

$$(x_1 + \bar{x}_3 + x_4)(\bar{x}_1 + \bar{x}_2 + \bar{x}_4)(\bar{x}_2 + x_3)(\bar{x}_1 + x_2 + x_4)$$

find an assignment of truth values (0 or 1) to the boolean variables x_1, x_2, x_3, x_4 so that the expression itself is true. This means that *each* of the clauses such as $x_1 + \bar{x}_3 + x_4$ must be true (have value 1). If we search for a satisfying assignment for the expression displayed above, we might try $x_1 = 1$, $x_2 = 1$, $x_3 = 1$, $x_4 = 1$, only to discover that the clause $\bar{x}_1 + \bar{x}_2 + \bar{x}_4$ has the value $0 + 0 + 0 = 0$ under this assignment, violating the condition that each of the clauses must be true. So it follows that this assignment does not satisfy the expression. However, when we try $x_1 = 0$, $x_2 = 1$, $x_3 = 1$, $x_4 = 1$, we discover that all clauses are true under this assignment and it "satisfies" the expression. Expressions having the form above are said to be in *product-of-sum* form.

Another way of viewing the satisfiability problem is to examine the circuit corresponding to the product-of-sums expression and to ask what combination of input values will cause the circuit as a whole to output a value of 1 (Figure 34.1).

Lest the preceding example give the impression that satisfiability problems are always easy to solve, consider this:

$$(x_1 + \bar{x}_2 + x_3)(\bar{x}_1 + \bar{x}_2 + \bar{x}_3)(\bar{x}_1 + \bar{x}_2 + x_3)(\bar{x}_1 + x_2 + \bar{x}_3)$$
$$(x_1 + x_2 + x_3)(\bar{x}_1 + x_2 + x_3)(x_1 + x_2 + \bar{x}_3)$$

This product-of-sums expression has more clauses than the previous one but fewer variables. One might try a number of assignments before arriving at

$$x_1 = 0 \qquad x_2 = 1 \qquad x_3 = 1$$

In fact, this is the only assignment that satisfies the expression. It is even possible for a product-of-sums expression to have *no* satisfying assignments, and in that case it is called *unsatisfiable*.

The obvious way of solving algorithmically a satisfiability problem with n logical variables is to generate all possible assignments systematically and to test each one with the given expression. If it satisfies the expression, quit; otherwise, continue generating and testing. If none of the 2^n possible assignments satisfies the expression, it is unsatisfiable. Unfortunately, the time taken by this algorithm tends to grow exponentially with n; with satisfiable instances of the problem, the algorithm may generate anywhere between 1 and 2^n assign-

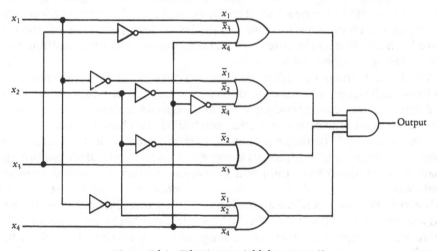

Figure 34.1 What inputs yield the output 1?

ments before arriving at a solution. With unsatisfiable instances, it must try all 2^n assignments, an unacceptable wait for all but the smallest problem instances.

A slightly better algorithm was discovered by M. Davis and H. Putnam in the early 1960s. In its simplest form, the Davis-Putnam algorithm could be presented as follows:

DAVIS – PUTNAM

procedure *split* (E)
    1. **if** E has an empty clause, **then** return
    2. **if** E has no clauses, **then** exit with current partial assignment
    3. select next unassigned variable, x_i, in E
    4. **split** $(E(x_i = 0))$
    5. **split** $(E(x_i = 1))$

Initially, E is the given product-of-sums expression for which a satisfying assignment is sought. At each step of the recursion, E represents the expression at that step. If the current partial assignment has failed to satisfy one of the clauses or has satisfied them all, the algorithm returns or exits, respectively. Otherwise, the current expression is scanned from left to right, and the first variable that appears, say x_i, becomes the basis for two more calls to **split**. For the first, the expression $E(x_i = 0)$ is formed. Each clause containing $\bar{x}_i$ is satisfied by $x_i = 0$ and therefore deleted from the expression, and each appearance of x_i is deleted from any clause in which it appears. For the second of the calls to split, $E(x_i = 1)$ is formed. Clauses containing x_i and appearances of $\bar{x}_i$ are deleted from E.

If at any time an expression containing an empty clause is generated, then that clause has failed to be satisfied by the current partial assignment and there is no point in continuing. If, however, all clauses have been deleted, then all have been satisfied by the current partial assignment, and any remaining variables may be assigned arbitrary values.

The Davis-Putnam algorithm, being a recursive procedure that calls itself twice at each stage, explores an implicit search tree for each expression inputted to it. The tree corresponding to the first example is shown in Figure 34.2. Each ∅ represents an expression from which all clauses have been deleted and hence a successful (partial) assignment. Each appearance of (∅) indicates an empty clause and a failed (partial) assignment. In this particular example, the Davis-Putnam algorithm would find a satisfying assignment almost immediately with $x_1 = 0$, $x_2 = 0$, $x_3 = 0$. The last variable, x_4, could be set to 0 or 1 arbitrarily. The success of the algorithm in this instance, however, has more to do with the relative abundance of solutions than any other factor. In general, the Davis-Putnam algorithm *may* take a long time with certain expressions but usually finds a solution in considerably less time than the exhaustive search

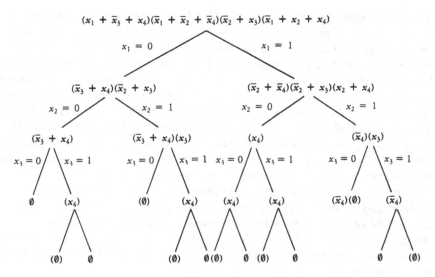

$$(x_1 + \bar{x}_3 + x_4)(\bar{x}_1 + \bar{x}_2 + \bar{x}_4)(\bar{x}_2 + x_3)(\bar{x}_1 + x_2 + x_4)$$

Figure 34.2 The tree implicit in the Davis-Putnam algorithm

algorithm mentioned earlier. The reason for this lies in the Davis-Putnam algorithm's ability to prune unsuccessful branches from its search tree.

There is no algorithm known at present that is guaranteed to solve an n-variable satisfiability problem in less than exponential time. Specifically, there is no algorithm which is guaranteed to solve it in polynomial time (see Chapter 15), say in time proportional to n^k or less for some fixed power k. Is this inability to find an efficient algorithm for the satisfiability problem due to our own relative stupidity, or is it just possible that no such algorithm exists?

To make matters worse, there is a much simpler version of this problem which seems to be equally difficult: Find an efficient (say, polynomial-time) algorithm which, for each product-of-sums expression, merely outputs a 1 if the expression is satisfiable and a 0 if it is not. This is called the *satisfiability decision problem* (or, sometimes, just the *satisfiability problem* when the context is clear). No satisfying assignment is asked for, just a decision about whether the expression is satisfiable.

To see why the satisfiability problem is central (and difficult), it is necessary to examine product-of-sums expressions in a new light. Rather than view them as logical expressions which are either true or false depending on how their variables are assigned, we may regard such formulas as a kind of language in which a great many mathematical ideas may be expressed. For example, consider this well-known problem in graph theory: Given a graph G, color its vertices red, yellow, and blue in such a way that if any two vertices are joined by an edge, then the vertices receive different colors.

233

Shown in Figure 34.3 are two graphs G and H. One is "three-colorable," and the other is not. One may, for example, begin to color either graph in the manner indicated, but sooner or later in one of them all attempts to extend this use of three colors to the entire graph will fail. Among the graphs which *can* be labeled with three colors, some are easy to color, and others are quite difficult. As a matter of fact, no one has succeeded in finding an efficient algorithm for the three-color problem either. Is there an algorithm which, for any graph G having n vertices, will find a way to use three colors (if it exists) for G in no more than n^k steps? One can even ask for an efficient algorithm for the corresponding problem: Is there an algorithm which, for any graph G having n vertices, will decide whether G is three-colorable in no more than n^k steps?

It turns out that the difficulty of the three-color problem for graphs is intimately related to the satisfiability problem for logical expressions. This is due to the potential of such expressions for expressing mathematical ideas.

There is an algorithm, shown below, which takes an arbitrary graph G as input and outputs a product-of-sums expression $E(G)$. This algorithm has two important properties:

1. If G has n vertices, then the algorithm requires no more than n^2 steps to output $E(G)$.
2. Graph G is three-colorable if and only if $E(G)$ is satisfiable.

Before displaying the algorithm, we remark at once that its very existence implies that the satisfiability decision problem is just as hard as (if not harder than) the three-color decision problem. For if we should find a polynomial-time algorithm for the satisfiability decision problem, it would only be necessary to couple it with the algorithm below to turn it into an efficient algorithm for the three-color decision problem: The composition of two polynomial-time algorithms is still a polynomial-time algorithm, and a yes/no answer for the first problem is tantamount to a yes/no answer for the second one.

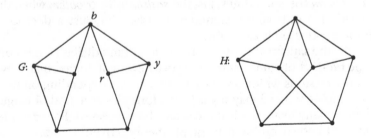

Figure 34.3 Which graph is three-colorable?

procedure *transform*
    for each vertex v_i in G **output**$(r_i + y_i + b_i)$
    for each edge (v_i, v_j) in G **output** $(\bar{r}_i + \bar{r}_j) \cdot (\bar{y}_i + \bar{y}_j) \cdot (\bar{b}_i + \bar{b}_j)$

Assuming that the string of clauses outputted by the algorithm is given the correct form, namely, that the clauses are written out as a product, it is not hard to establish the assertion 2 above. However, if G is three-colorable, then we may assign values to the logical variables r_i, y_i, and b_i as follows: Let "color" be a three-color arrangement of G, and define

$$r_i = \begin{cases} 1 & \text{if color } (v_i) = \text{red} \\ 0 & \text{otherwise} \end{cases}$$

$$y_i = \begin{cases} 1 & \text{if color } (v_i) = \text{yellow} \\ 0 & \text{otherwise} \end{cases}$$

$$b_i = \begin{cases} 1 & \text{if color } (v_i) = \text{blue} \\ 0 & \text{otherwise} \end{cases}$$

Certainly, each vertex will receive one color, so one of the variables in $r_i + y_i + b_i$ is true. At the same time, whenever v_i and v_j are joined by an edge, they receive different colors. They cannot both be red, so one of $\bar{r}_i$ or $\bar{r}_j$ must be true in $\bar{r}_i + \bar{r}_j$. Similarly, $\bar{y}_i + \bar{y}_j$ and $\bar{b}_i + \bar{b}_j$ are satisfied. It is only slightly more difficult to show that if one is given a satisfying assignment for the product-of-sums expression produced by the algorithm, then one may obtain a three-color arrangment for G from it. Indeed, because the latter solution may be obtained from the former very quickly, what we have said about the relative difficulty of the decision problems applies with equal force to the more general problems: Algorithmically speaking, it is at least as hard to find a solution to the satisfiability problem (when it exists) as it is to find a solution to the three-color problem.

There are a great many problems like the three-color one. Each can be transformed to the satisfiability problem, and each one seems to be very difficult in that no polynomial-time solution algorithm for the problem is known to exist (see Chapter 41). In this sense, satisfiability is a "central" problem, and in this same sense satisfiability is a very hard problem, indeed. It might be reasonable to suspect rather strongly that a fast algorithmic solution for the satisfiability problem simply does not exist.

Problems

1. Draw the implicit search tree for the second product-of-sums example in this chapter. Under what circumstances will the Davis-Putnam algorithm produce frequent failures high in its search tree? How can you improve the algorithm?

2. Write out the product-of-sums expressions generated by the transformation algorithm for the graphs G and H in Figure 34.3. Produce a satisfying assignment for one of these expressions based on your experience in attempting to find a three-color arrangement for the graphs.

3. The hamiltonian circuit problem for graphs is to find a circuit that passes through each vertex exactly once. Devise a polynomial-time algorithm which transforms an instance of this problem (in the form of a graph G) to an instance $F(G)$ of the satisfiability problem. Not only must the algorithm run in polynomial time (in n, the number of vertices), but also G must have a hamiltonian circuit if and only if $F(G)$ is satisfiable.

References

M. R. Garey and D. S. Johnson. *Computers and Intractability*. Freeman, San Francisco, 1979.

Chin-Liang Chang and Richard Char-Tung Lee, *Symbolic Logic and Mechanical Theorem Proving*, Academic Press, Bethesda, Md., 1973.

35

SEQUENTIAL SORTING

A Lower Bound on Speed

S orting is a computational task with enormous commercial, industrial, and scientific implications. Many commericial and institutional computers spend a significant fraction of their time sorting lists. So it is very practical to ask,

> Given a list of n elements, what is the minimum
> amount of time a computer may take to sort it?

Elsewhere in this book there is an algorithm which can accomplish this task in time $O(n \log n)$ (see Chapter 40). In other words, there is a constant c so that for large enough n, the number of steps this algorithm requires to sort n elements is never greater than

$$c \cdot n \log n$$

It turns out that this quantity has also the right order of magnitude as a *lower bound* on the number of steps required, at least when the question above is appropriately refined.

If "sorting" is restricted to mean that the computer will compare two ele-

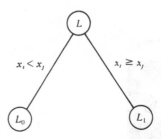

Figure 35.1 Initial decision of a sorting algorithm

ments on the n-element list at one time, then this kind of lower bound is not at all difficult to demonstrate. Let the list to be sorted be represented as

$$L = x_1, x_2, \ldots, x_n$$

The computer begins by comparing x_i with x_j, let us say, and does one thing or another as a result of the outcome (Figure 35.1). What the computer does is alter the list to L_0 or L_1 depending on this comparison; for example, it may exchange or not exchange x_i and x_j in creating L_0 or L_1. In any event, depending on circumstances, it then goes to work on either L_0 or L_1, where precisely the same sort of observation applies. Continuing this process results in a tree whose bottom nodes all correspond to potential sorted versions of the input list.

For example, if L has just three elements, Figure 35.2 shows how the "decision tree" might appear.

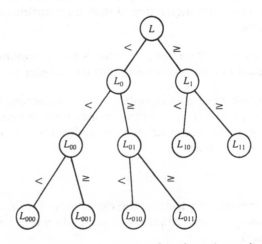

Figure 35.2 The decision tree for a three-element list

To understand the full significance of the terminal nodes, it is useful to imagine what the sorting algorithm does with all $n!$ possible list versions of an n-element set. Given any two such lists, suppose that the algorithm took the same branch for both lists at each level of the tree, so that the same terminal node was arrived at in each case.

Each time the sorting algorithm makes a decision as the result of a single comparison, it carries out a permutation π on the current version of input list L. Moreover, it performs this permutation on whatever list it currently has. Thus if two distinct list versions of the same set end up at the same terminal node, it follows that they have been permutated identically. But this means that at most one of these lists can have been sorted correctly. The only conclusion to be drawn from all this is that if the sorting algorithm performs correctly, the corresponding decision tree must have at least $n!$ terminal nodes.

Now a binary tree with m terminal nodes must have a depth of at least $\lceil \log m \rceil$, and the depth of our decision tree must therefore satisfy

$$
\begin{aligned}
D &\geq \lceil \log n! \rceil \\
&> \log n(n-1)(n-2) \cdots n/2 \\
&> \log \left(\frac{n}{2} \right)^{n/2} \\
&= \frac{n}{2} \log \frac{n}{2} \\
&= O(n \log n)
\end{aligned}
$$

This is just another way of saying that any comparison sort taking an n-item list as input must perform at least $O(n \log n)$ comparisons in the worst case. Any algorithm that performs substantially better than this, overall, must be making its sorting decision on some other basis or must be capable of making many comparisons in parallel (see Chapter 62).

Problems

1. A bubble sort operates on an n-element input list by scanning the list until it finds the first element that is smaller than its predecessor. The algorithm exchanges the smaller element with each predessor larger than it until the element comes to rest in its proper place. The algorithm then picks up scanning where it previously left off. What is the worst-case complexity of this algorithm? Does it realize the lower bound of this chapter in order of magnitude?

2. Draw the decision tree for the bubble sort algorithm operating on a three-element list.

References

A. V. Aho, J. E. Hopcroft, and J. D. Ullman. *The Design and Analysis of Computer Algorithms.* Addison-Wesley, Reading, Mass., 1974.

Donald E. Knuth. *The Art of Computer Programming,* vol. III. Addison-Wesley, Reading, Mass., 1967.

36

NEURAL NETWORKS
THAT LEARN

Converting Coordinates

N eural networks, assemblies of artificial neurons that seem capable of certain perceptual and cognitive tasks, captured the imagination of computer scientists in the 1980s when it turned out that two- and three-layer networks could recognize patterns that were beyond the capabilities of one-layer networks (see Chapter 27). Such networks also seemed capable of learning a wide variety of tasks. Although relatively few computer scientists study neural nets and their learning abilities, several hundred scientists have entered the field, mainly from engineering, physics, and even psychology.

This chapter explains the general paradigm, then describes a simple network that readers may construct for themselves in a computer. The network, simulated by software, learns how to convert polar coordinates to Cartesian coordinates!

Most learning networks currently under study share the basic architecture shown in Figure 36.1. Information enters the network on the left through a layer

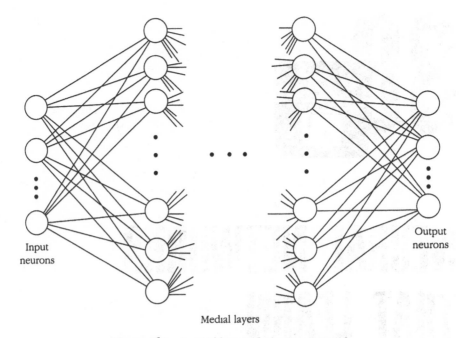

Figure 36.1 General layout of a learning network

of *input neurons,* enters one or more layers of *medial neurons,* then departs, on the right, through a layer of special *output neurons.* The network processes the information as it flows through successive layers, from input to output. Each neuron in a given layer communicates with every neuron in the next layer via special synaptic connections. In formal terms, a *synapse* amounts to a multiplier, or weight, which modifies the real number that is transmitted from one neuron to the other.

Individual neurons operate, like their biological counterparts, by a process of summation. Each neuron adds together all the signals that it receives, then transmits the sum to all neurons in the next layer. If it happens to reside in one of the medial layers, it will modify the sum of its signals in a special way before transmitting it. A special, sigmoidal function squeezes the signal nonlinearly into the interval from -1 to $+1$: The larger a sum is, the more closely it will approach $+1$ or -1, depending on its sign. Figure 36.2 displays one sigmoidal function.

Sigmoidal functions enable neural networks to respond nonlinearly to their environments. A great variety of actual functions, with names like hyperbolic tangent, arc tangent, Fermi function, and so on, play the role quite nicely. Their ability to keep the outputs of intermediate neurons bounded (between 0 and 1) is just as important as their nonlinear response.

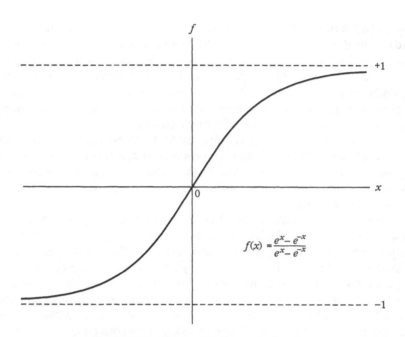

Figure 36.2 The hyperbolic tangent function

When such a network has been trained and is ready to run, input signals are fed to the input neurons and the network does the rest. If the first medial layer contains 30 medial neurons, for example, the signal from each input neuron would automatically divide into 30 separate signals, such a real number, each destined for one of the 30 medial neurons via a synapse. The synapse multiplies the signal by a special real number called the *synaptic weight*. Each medial neuron adds together the signals it receives from the input neurons (via the synapses), applies its sigmoidal function, then transmits the number (now a real number between -1 and $+1$) to the next layer of medial neurons or to output neurons if it happens to be in the last medial layer. On the way, each signal must again pass through a synapse where it is modified by the weight peculiar to that synapse.

Neither input neurons nor output neurons apply a sigmoidal function to the signals they receive. Input neurons merely transmit the signal they receive while output neurons simply add them together.

In addition to a number of regular input neurons, a neural net may also include a *bias neuron*. This special neuron regulates the operation by ensuring that no neuron receives a zero signal. It also gives experimenters an additional degree of freedom, enabling them to shift all the input signals by a new, constant amount.

The network that computes Cartesian coordinates from polar coordinates is based on one developed by Edward A. Rietman and Robert C. Frye of AT&T Bell Laboratories in the late 1980s as a demonstration of network learning techniques. The program converts polar coordinates to Cartesian coordinates by thousands of trials and errors, adjusting its behavior as it goes. The program has two parts, one that does the conversion and one that modifies the synaptic weights to improve the accuracy of future conversions.

The Rietman–Frye network requires only 30 to 40 neurons (in a single medial layer) to learn coordinate conversions. It has two input neurons that carry the numerical signals representing the polar coordinates to be converted. At the other end of the network, two output neurons represent the corresponding Cartesian coordinates.

The Rietman–Frye network employs a common learning algorithm called the *back propagation technique.* In what follows, readers may wish to consult the algorithmic outline on page 247 now and then. Two arrays, called *synone* and *syntwo,* contain all the synaptic weights. The first array consists of the synapses between the input neurons and the medial ones. The second array consists of the synapses connecting the medial neurons to the output neurons. The net "learns" when it changes the synaptic weights in these arrays as a result of its experience with the coordinate pairs that it encounters.

In response to a particular set of polar coordinates, the network will develop two error differences, *e1* and *e2,* between the target coordinates and the computed ones: These differences form the basis for adjustments in the synaptic weights all the way through the net, from back to front, hence the term "back propagation."

In the Rietman–Frye network, the method first calculates how the weights in the second set of synapses must be changed to reduce the error if the same conversion were to be attempted again. To adjust the synaptic connection between the ith medial neuron and the jth output neuron, for example, POLARNET changes *syntwo(i, j)* by adding to it the product of the jth error and the previous output of the ith medial neuron.

To avoid problems of overshoot, an additional multiplier must enter the adjustment formula. A learning rate parameter called *rate,* usually with a value somewhere in the neighborhood of 0.1, modifies the adjustment to a gentler level.

Back propagation next alters the values of the first set of weights, contained in the array *synone,* by essentially the same method. First, however, the derivative of the sigmoid function must be applied to the back-propagated sum of the adjusted synapse values in *syntwo.* This step assists in the error minimization process by ensuring that, in a statistical sense, the first set of synapses will adjust in a direction that reduces the error on subsequent conversions.

For each medial neuron, the back-propagation procedure forms the product between each of the two error terms and their corresponding synapse values for that particular medial neuron. It adds the two products together and then, pretending that this was the output of the medial neuron in question, computes the corresponding error-improving input by using the derivative of the sigmoidal function, in this case, $1 - y^2$.

The weights stored in *synone* can now be changed in the final phase of back propagation. The actual adjustment, called *delta,* is the product of four numbers: *rate,* the rate as described earlier, *sigmoid,* the reconstructed output from the medial neurons, *sigma,* the error-reducing signal that propagates backward from the medial neurons, and *input,* the input to the neuron at the front end of the synaptic connection. The actual formula appears in the second algorithm (see page 247).

The standard error, *E,* combines the two individual error terms (the square root of the sum of the squares of *e1* and *e2*) to measure the network's accuracy at any given time. Figure 36.3 shows how *E* changed in an automated experiment involving 50,000 randomly selected coordinate conversion examples, each involving a point within one unit distance of the origin.

This particular network had 40 medial neurons and a rather modest learning *rate* of 0.01. The learning curve, in spite of frequent fluctuations, shows a

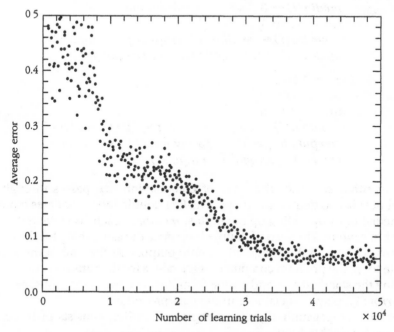

Figure 36.3 The learning curve of the network

determined downward trend from 50% down to about 6%. It reaches this level of competance shortly after 30,000 trials and does not improve much thereafter. Networks with two medial layers may do better. For example, a network having two medial layers of 15 neurons each converged to a competence level of only 1.5% error by the 25,000th test. Strangely enough, in the world of neural nets, bigger is not necessarily better. The network's performance becomes distinctly worse if it has 150 neurons in each medial layer instead of 15!

The foregoing descriptions of the coordinate-conversion network will enable readers with a bit of programming experience to write POLARNET using the following algorithms as a guide. But before the program carries out any conversions, it must initialize the values of *synone* and *syntwo*, setting each to to $0.1 * rand$, for a succession of standard random numbers, *rand*. Such small and variable values for the synapse weights are crucial to get the network running. If the weights were all the same to begin with, they would stay the same throughout training and the network would never change its behavior in a meaningful way.

Part one: Coordinate conversion

```
input input(1), input(2)
for i ← 1 to n
    medin(i) ← 0              / medial input
    for j ← 1 to 3
        medin(i) ← medin(i) + synone(j, i) * input(j)
    medout(i) ← hyperbolic tangent (medin(i))

for i ← 1 to 2
    output(i) ← 0
    for j ← 1 to n
        output(i) ← output(i) + syntwo(j, i) * medout(j)
    compute target(i)      /using input(1) and input(2)
    error(i) ← target(i) − output(i)
```

The algorithm simulates the wave of computation that passes through the neural net via two double **for** loops. The first double loop calculates inputs to the medial layer by adding up the inputs, *medin*, to each, as weighted by the entries in *synone*. The second double loop does the same thing for the output neurons, adding up the weighted *medout* outputs. At the end of the second double loop, the program computes target values for the cartesian coordinates by using the standard sine and cosine formulas. The two error terms, *error(1)* and *error(2)*, play a crucial role in the next procedure.

The back-propagation algorithm, explained earlier, consists of three sections, each a double loop.

Part two: Back propagation algorithm

> **for** $i \leftarrow 1$ **to** 2 /adjust the second synaptic layer
>   **for** $j \leftarrow 1$ **to** n
> $syntwo(j, i) \leftarrow syntwo(j, i) + rate * medout(j) * error(i)$
>
> **for** $i \leftarrow 1$ **to** n /derive the sigmoidal signal
> $sigma(i) \leftarrow 0$
>   **for** $j \leftarrow 1$ **to** 2
> $sigma(i) \leftarrow sigma(i) + error(j) * syntwo(i, j)$
> $sigmoid(i) \leftarrow 1 - (medin(i))^2$
>
> **for** $i \leftarrow 1$ **to** 2 /adjust the first synaptic layer
>   **for** $j \leftarrow 1$ **to** n
> $delta \leftarrow rate * sigmoid(j) * sigma(j) * input(i)$
> $synone(1, j) \leftarrow synone(i, j) + delta$

The program uses the two algorithms inside the following train-and-test loop:

> **1.** initialization
> **2. for** count $\leftarrow 1$ **to** 10,000
>   **1** select random point
>   **2** convert coordinates
>   **3** back propagation
>   **4 output** error E

Step 2.1 must generate random points from the unit circle. Step 2.4, **outputting** the error E, could mean anything from printing the value to plotting it on the screen. A plot of error values such as those in the learning curve in Figure 36.3 makes the program more informative and useful.

Readers may alter the program to taste. Not only may they change the total number of experiments (from 10,000, above), but they are also free to change the number of input neurons (3), output neurons (2), or medial neurons (n) to any number they like. They may even install two or more medial layers to see how much difference they make to the speed of learning. In such a case, however, both the forward computation and backward propagation procedures will be in for something of an overhaul.

It cannot be guaranteed that neural networks will converge to useful behavior in every potential application. Mathematically speaking, the weakness stems from the fact that neural net learning is essentially a hill-climbing (or valley-descending) algorithm, one that proceeds through its search space toward a local optimum, then stops.

Each possible combination of synaptic weights corresponds to a point in an m-dimensional space, where m is the total number of synapses. Each point in this space also has an average error value associated with it. These values constitute a vertical landscape, a terrain of ups and downs imposed on the m-dimensional map of all possible synaptic combinations. The question of whether neural nets can be effective at a given discrimination job really boils down to two questions:

a. Does a truly optimal valley exist in the first place?

b. Can the neural net find such an optimum before it settles into a suboptimal trough?

A "truly optimal valley" means a set of synapse values that enable the net to meet some previously imposed standard of minimal competance. Assuming that some optimal combination of synaptic values actually exists, the landscape of possible behaviors may have enough local optima that finding the global optimum combination is computationally hopeless from a complexity point of view (see Chapter 15) even for a parallel network (see Chapter 62).

Neural networks have found a small niche in our software toolkit. How much larger that niche becomes will depend, ultimately, on the range of tasks that they can reliably learn to perform.

Problems

1. Before it has been trained, nothing about the network suggested for coordinate conversion makes it especially suitable for that task alone. Try training the network on a completely different task: learning to add! Use only one of the output neurons for the sum and back propagate from that neuron alone.

2. Some neural nets do not use negative numbers. Their sigmoid functions will therefore resemble the Fermi function,

$$f(x) = (1 - e^{-x})^{-1}$$

Show that no such net will be able to act as a reliable comparator, distinguishing which of two inputs is the larger.

References

D. E. Rumelhart and J. L. McClelland (eds.). *Parallel Distributed Processing: Explorations in the Microstructure of Cognition, Vol. 1.* MIT Press, Cambridge, Mass., 1986.

E. A. Rietman. *Explorations in Parallel Processing.* Windcrest/Tab Books Inc., Blue Ridge Summit, Pa., 1990.

37

PUBLIC KEY CRYPTOGRAPHY

Intractable Secrets

S ecret communication is a need not only of military commanders and intelligence agents but also of large commercial and industrial firms relaying confidential information. In today's world probably at least 90 percent of sensitive or secret information transmitted electronically arises in the latter sector. One has only to think of the flow of credit information among various financial institutions.

The tremendous quantities of such information make it imperative that confidentiality be guaranteed by a system of encryption which is both computationally feasible and yet very hard to decrypt when information is intercepted.

Traditionally, the encryption process has involved two transformation algorithms T and T^{-1} operating on a message. For example, in Figure 37.1 the message X might be an English sentence, and the transformation T might be the substitution, for each letter of the alphabet, of the kth letter beyond that one.

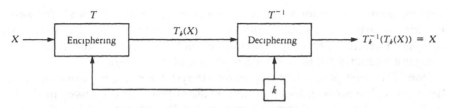

Figure 37.1 Traditional cipher systems

(When we reach the end of the alphabet, we add the blank character and then start all over again at A.) For example, with $k = 4$ the word SECRET becomes WIGVIX. If the receiver and sender of the message know both the algorithm and the key k, then enciphering and deciphering a message are both very easy operations, especially by computer. If the algorithm becomes public knowledge, however, it would not take a cryptanalyst long to discover the key k once a particular message had been intercepted: It is merely necessary to try $k = 1, 2, 3, \ldots$ until the inverse transformation (substituting the kth letter previous to the one being decrypted) yields an intelligible message. In this particular case, even if the encryption algorithm were not public, it would not take a cryptanalyst long to "crack the code." However, far more elaborate transformations of X are available, transformations which are so complicated that even if they were made public, knowledge of the key would be essential to deciphering the message. Thus traditional encryption systems still require that keys be distributed (usually by courier) to the various users of the system. With frequent key changes, this can be very costly and, in some cases, too costly and time-consuming even to consider.

In public key cryptography, each user of the system is given a key k which never needs to be changed and may even be listed in a public directory! The receiver of messages in this system (see Figure 37.2) however, has a private key k' which is used in implementing the inverse transformation T^{-1}. Naturally, there is a relationship between k and k', and in fact the nature of that relation-

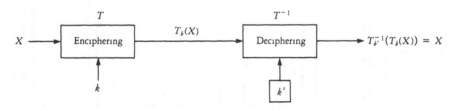

Figure 37.2 The public key system

ship can even be revealed without compromising the security of k'. The essential idea is that k' is awfully hard to compute, given k. But k is very easily computed given k', so easy that from k' one may, in a matter of moments, compute a unique public key for a new user of the system.

One of the first public key cryptographic systems was developed by Martin Hellman, Ralph Merkle, and Whitfield Diffie at Stanford University in the late 1970s. It was based on a certain computationally difficult problem called the *subset-sum problem*. In much of the literature on public key cryptography, this is called the *knapsack problem*, but it is really a special case of it.

> *Subset-Sum Problem:*   Given $n + 1$ positive integers $a_1, a_2, \ldots, a_n$ and B, find a subset of the a_i that sums to B.

For example, suppose we are given both $n = 5$ integers 5, 8, 4, 11, 6 and $B = 20$. Can we find a subset of the five integers whose sum is 20?

Another way of viewing this problem is to imagine that the n integers are the heights of blocks that fit snugly into a box of height B (Figure 37.3). Can one find a subset of the blocks which exactly fills the box (see Chapter 30)?

In this example, it does not take very long, after a few combinations are tried, to discover that $5 + 4 + 11 = 20$. In general, however, no one knows an algorithm that is guaranteed to solve the subset-sum problem in polynomial time. In fact, the problem is NP-complete (see Chapter 41). Thus any algorithm guaranteed ultimately to solve the problem will probably have to try a significant fraction of the 2^n possible combinations of integers to discover which subsets (if any) sum to B.

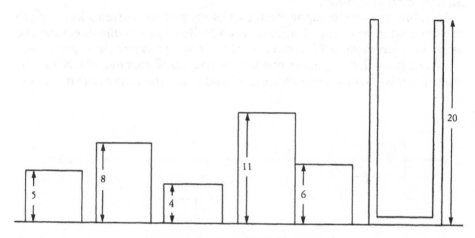

Figure 37.3   The subset-sum problem

Here is how the cryptographic system based on this problem works: Suppose a user has some confidential information to transmit to the receiver. The user has been given a public key consisting of n integers $a_1, a_2, \ldots , a_n$. The information is transformed to a string of binary digits, and then these are partitioned into blocks of length n. The block $x = (x_1, x_2, \ldots , x_n)$ is mapped into the integer

$$B_x = \sum_{i=1}^{n} x_i a_i$$

and transmitted.

To make the encryption process more concrete, suppose that the message is SECRET and that it is encoded in 7-bit ASCII:

S	E	C	R	E	T
1010011	1000101	1000011	1010010	1000101	1010100

For convenience in this example, we take $n = 7$ (although much higher values are normally used) and thereby encipher the first letter S as

$1 \times (901) + 0 \times (568) + 1 \times (803) + 0 \times (39)$
$\qquad\qquad\qquad\qquad + 0 \times (450) + 1 \times (645) + 1 \times (1173)$

where (901, 568, 803, 39, 450, 645, 1173) is the public key $(a_1, a_2, \ldots , a_n)$ and (1,0,1,0,0,1,1) is the block x being enciphered. The resulting integer $B_x = 3522$ is then transmitted over the insecure channel. Anyone wishing to know what message is encrypted in 3522, even if he or she knew that some combination of the public key integers was involved, might still be trying a significant fraction of the $2^7 = 128$ possible combinations before arriving at the correct one. If n is very much larger than 7, the time it might take to do this would become prohibitive. Why devote 100 years of computing time to deciphering a piece of information which, though secret now, will almost certainly be irrelevant by the time it is deciphered? In fact, with very little additional increase in n, an interloper might be faced with a problem which would take the lifetime of the universe to solve.

When the message B_x arrives at its intended destination, the receiver uses a private key $(a'_1, a'_2, \ldots , a'_n)$ and two special integers w and m to decipher the message.

The public key $(a_1, a_2, \ldots , a_n)$ was originally computed on the basis of this private information as follows:

$$a_i = w \cdot a'_i \bmod m$$

To retrieve the message bits x_i, the receiver formulates a special version of the subset-sum problem:

Which subset of $(a'_1, a'_2, \ldots, a'_n)$ sums to B'_x, where

$$B'_x = B_x \cdot w^{-1} \bmod m$$

Here, w^{-1} is the inverse of w in the field of integers modulo m. In other words, $w \cdot w^{-1} \equiv 1 \bmod m$.

The solution of the receiver's special subset-sum problem is really the inverse transformation T^{-1}. Thus, if the receiver's private key happens to be $(1, 2, 5, 11, 32, 87, 141)$, along with the special integers $w = 901$ and $m = 1234$, then the problem is to discover which subset of the private key integers sums to

$$\begin{aligned} B'_x &= 3522 \times (901)^{-1} \bmod 1234 \\ &= 3522 \times 1171 \bmod 1234 \\ &= 234 \end{aligned}$$

This problem happens to be very easy to solve quickly. The receiver's private integers are arranged so that each is larger than the sum of the integers preceding it in the sequence. It is not hard to show that the following algorithm solves this problem in linear time:

SUMSOLVE

1. $sum \leftarrow 0$
2. for $i = n$ step -1 until 1 do
       if $a_i + sum \leq B'_x$
          then $sum \leftarrow sum + a_i$;
             $subset(i) \leftarrow 1$
          else $subset(i) \leftarrow 0$
3. if $sum = B'_x$ then exit with subset
            else exit with "failure"

This algorithm starts with the largest (rightmost) integers in the private key, attempting to fit them into the box, so to speak, and discarding any which do not fit. It works because each a_i is larger than the sum of the integers which will follow it in this right-to-left order.

Applying SUMSOLVE to the instance $(1, 2, 5, 11, 32, 87, 141)$ and $B'_x = 234$, we quickly find that subset $= (1,0,1,0,0,1,1)$. Of course, this is the ASCII string 1010011 for S, the first letter of the SECRET message.

Why did everything work out so nicely? Perhaps the algebra of T and T^{-1} provides the simplest explanation:

$$B_x = \sum_{i=1}^{n} x_i a_i$$

$$= \sum_{i=1}^{n} x_i \cdot wa'_i \bmod m$$

but

$$B'_x = B_x \cdot w^{-1} \bmod m$$

$$= \sum_{i=1}^{n} x_i (wa'_i \bmod m)(w^{-1} \bmod m)$$

$$= \sum_{i=1}^{n} x_i a'_i$$

In other words, the message x encoded in B_x as $\sum_{i=1}^{n} x_i a_i$ turns out to be encoded in B'_x as $\sum_{i=1}^{n} x_i a'_i$.

Public key codes have only two possible points of vulnerability (short of the private key being discovered):

An algorithm which solves NP-complete problems quickly
An algorithm which solves the particular (NP-complete?) problem on which a cryptographic system is based.

The first possibility is considered to be highly unlikely by experts in complexity theory, but the second can happen if the particular problem does not have adequate theoretical safeguards. This is precisely what happened to the Hellman-Merkle-Diffie cryptographic system just described. It turned out that although the "public" subset-sum problem appeared to be a general version, including those cases which presumably make the problem computationally intractable, it really was not. Adi Shamir, at the Massachusetts Institute of Technology, discovered that it was much more of a special case (rather like the "private" subset-sum problem) than its designers realized. Shamir found a polynomial-time algorithm that solved even the public version of the subset-sum problem.

In the 1980s, a new public key cryptographic system supplanted the Hellman-Merkle-Diffie scheme. Called the *RSA cryptosystem* (short for Rivest, Shamir, and Adleman, three U.S. computer scientists who proposed it in the late 1970s), it depends on the apparent polynomial-time unsolvability of the factorization problem. If an *n*-bit number is not prime, how long must it take to find a nontrivial factor by computer? If, as many scientists think, it cannot be solved in a polynomial number of steps, then the following cryptosystem has a chance of succeeding.

In the RSA cryptosystem, two integers *e* and *n* are given as public keys. If a message *m* is converted to an integer less than *n*, then *m* may be encrypted according to the formula

$$c = m^e \bmod n$$

The intended receiver has a private key that consists of two prime factors, *p* and *q*, of *n*. In other words, $n = pq$. This means that the receiver of the message can easily find an integer *d* such that

$$ed = 1 \bmod \phi(n)$$

where $\phi(n) = (p-1)(q-1)$. To decipher message *c*, the receiver computes $c^d \bmod n$. This number can be analyzed as follows:

$$
\begin{aligned}
c^d \bmod n &= m^{ed} \bmod n \\
&= m^{k\phi(n)+1} \bmod n && \text{for some integer } k \\
&= m \cdot m^{k\phi(n)} \bmod n \\
&= m \bmod n && \text{since } m^{\phi(n)} = 1 \bmod n \\
&= m && \text{since } m < n
\end{aligned}
$$

Since *n* can be chosen with arbitrary factors *p* and *q*, the factoring problem exploited in the RSA cryptosystem is completely general and not open to a special attack, as was the previous scheme. Efforts continue in the direction of developing fast factorization algorithms. So far, the fastest such algorithm requires

$$O(e^{(\log n \, \log\log n)^{1/2}}) \text{ steps}$$

The whole question of the impact of theoretical investigations on public key cryptosystems has made certain authorities who are already dependent on such systems understandably nervous. In 1981 the U.S. National Security Agency requested a study of the matter.

Problems

1. Suppose the receiver's private key happens to be (2, 5, 18, 26, 82, 135, 280). If $w = 1003$ and $m = 1209$, work through the enciphering/deciphering procedure with S.

2. Devise a new public key encryption system based on the partition problem. Under what circumstances can the pseudo-polynomial time algorithm presented in Chapter 30 be used to break this system?

3. Show that $m^{Q(n)} = 1 \bmod n$ for any integer m less than $n = p \cdot q$.

References

Bruce Bosworth, *Codes, Ciphers, and Computers: An Introduction to Information Security.* Hayden, Rochelle Park, N.J., 1980.

Dorothy Elizabeth Robling Denning, *Cryptography and Data Security.* Addison-Wesley, Reading, Mass., 1983.

38

SEQUENTIAL CIRCUITS

A Computer Memory

In the field of logic design, there is a fundamental distinction between combinational circuits and sequential circuits (Figure 38.1). The former can be represented by a simple black box with input lines on one side and output lines on the other. Inside the black box is a logic circuit (see Chapter 3) composed of gates; every logical path in the circuit leads from an input to an output, and there are no circular logical paths. The latter kind of circuit contains two-state memory elements which can store or "remember" that state over a period of time.

In a combinational circuit a given pattern will always give rise to the same output, but in a sequential circuit this is no longer true: The current content of memory forms an additional input to the logic section, and the output may change from one occasion to the next, even with the same input. The ability to modify its operation with time is a distinguishing feature of modern computers. Time itself is marked off into discrete steps by pulses emanating from a clock. These pulses coordinate activity within a computer and within the somewhat humbler sequential circuits described here.

The basic unit of memory could be called a *cell*. There are various ways of

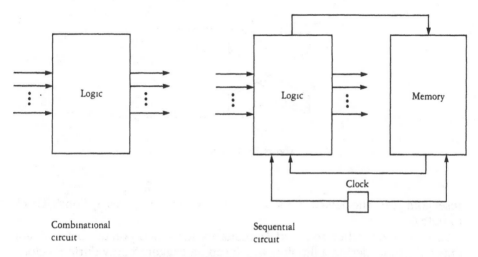

Combinational
circuit

Sequential
circuit

Figure 38.1   Combinational and sequential circuits

designing cells from simple logic gates, but these are used largely for purposes of illustration and do not correspond to memory cells in actual computers — except in function. An actual cell is generally fabricated in silicon technology from a collection of transistors and other electronic components (see Chapter 56).

We follow the former approach, maintaining the pleasant fiction that all the components of devices considered are made from basic gates such as AND, OR, and NOT gates or NAND gates and inverters. In fact, this does no real harm since at the level of hardware design it is easy to translate a specification in gates to one involving transistors. Indeed, some kinds of transistors turn out to be gates in any event.

A very simple memory cell called an *RS* flip-flop can be made from two NAND gates, as shown in Figure 38.2. This simple circuit is not combinational by our earlier definition since it contains a circular logical path (shown by a dashed line). The output of each NAND gate is sent to the input of the other. This makes the analysis of this circuit a bit tricky. Rather than start at the input side, we start at the output side. Can both Q and Q' be 0? Evidently not, since a NAND gate must output a 1 if either input is a 0. Care must therefore be taken to ensure that R and S are never simultaneously 0. Assuming, then, that only two states are possible, namely, $Q = 1$, $Q' = 0$ (labeled 1) and $Q = 0$, $Q' = 1$ (labeled 0), Figure 38.3 expresses the operation of the *RS* flip-flop as a finite automaton (see Chapter 2). The labels on the arcs encode the inputs as a 2-bit binary number *RS*. When the flip-flop is in state 0, if the set bit *S* goes to 1 simultaneously with the

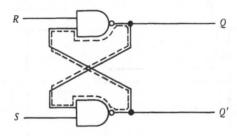

Figure 38.2 A flip-flop

reset going to 0, the device "flips" into state 1. From there, a reset "flops" it back to state 0.

As mentioned earlier, sequential circuits normally incorporate clocks. We are thus obliged to design a flip-flop which can be triggered only during a clock pulse. The circuit in Figure 38.4, incorporating the circuit above, accomplishes this.

The clock pulse (which we assume arrives as a 1) is said to *enable* either a set S or reset R signal because only when the two OR gates in Figure 38.4 receive a 0 (inverted 1) from the clock can either of the other two signals get through. The set signal then drives the device into state 1 if it is not already in that state. This tells the flip-flop, in effect, to "remember 1." The reset signal, similarly, tells it to "remember 0." If we represent the circuit above by a black box with appropriately labeled inputs and outputs, then we can design various forms of memory besides the cells mentioned earlier. For example, every computer contains a number of "working registers" through which the flow of information is much more frequent, on average, than the traffic in a specific cell of main memory. Figure 38.5 shows the first two bits of an n-bit register.

This register is organized vertically. Each box represents a flip-flop of the type just studied which can change state only when a clock pulse arrives. The individual bits currently being remembered in the register's flip-flop comprise its numerical content. This content can also be changed only when the LOAD

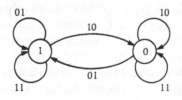

Figure 38.3 The flip-flop as an automaton

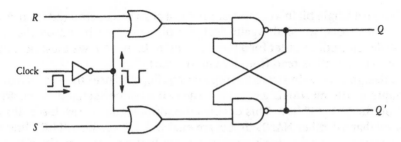

Figure 38.4 A clocked flip-flop

line carries a 1. This causes the outputs of the twin NAND gates feeding a single flip-flop, both normally 1, to transmit an inverted form of the signals arriving along the x_1 line and its inversion. Thus, if $x_1 = 1$ and the LOAD signal is 1, the R input will be 0 while the S input is a 1. This sets the flip-flop if it is not already set.

In this manner, pulses arriving simultaneously on input lines $x_1, x_2, \ldots, x_n$ can be loaded in parallel into the register. The output lines reflect the register's current content and can be read at any time. The register even has a CLEAR line which has the function of setting all the bits to 0.

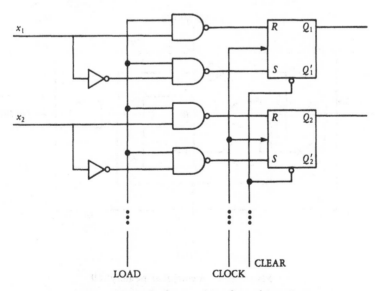

Figure 38.5 The first two bits of an n-bit register

Although a single bit in a computer's memory can be constructed from as few as two transistors, we shall again design a memory cell based on the logic already developed. In order for a memory cell to be useful, we must be able to write into it as well as read from it (Figure 38.6).

This design is very similar to the register flip-flop and uses the same enabling technique involving NAND gates. First, the cell must be selected by the arrival of a 1 on the SELECT line. This enables the input signal to reach the flip-flop provided that the other NAND inputs are enabled. A 1 on the SELECT line also helps to transmit the flip-flop's current content to the outside world. When the computer intends to write something in this memory cell, it not only turns on the SELECT line but also sends a 0 along the READ/WRITE line. This signal is inverted before it reaches the two NAND gates and so enables them in concert with the other 1 on the SELECT line. A 1 on the READ/WRITE line, however, disables the inputs to the flip-flop and enables the output. Thus a 1 means READ and has that function.

Representing such a memory cell by a single box, we can now design a simple memory of eight 4-bit words (Figure 38.7). Only the first two and the last of the eight words are shown below as rows of four cells each. Here we visualize a set of four inputs arriving at memory along parallel lines. The SELECT bit of a particular row of cells is turned on by the decoder. As described in Chapter 28, this device takes a small number of bits as input and outputs a 1 along the line whose number is encoded by the input. Three bits are sufficient to encode any

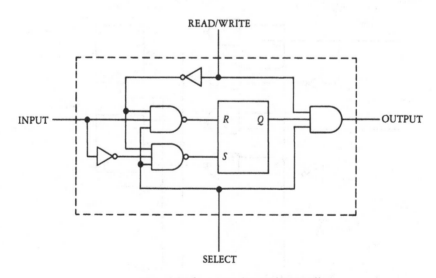

Figure 38.6 A complete memory cell

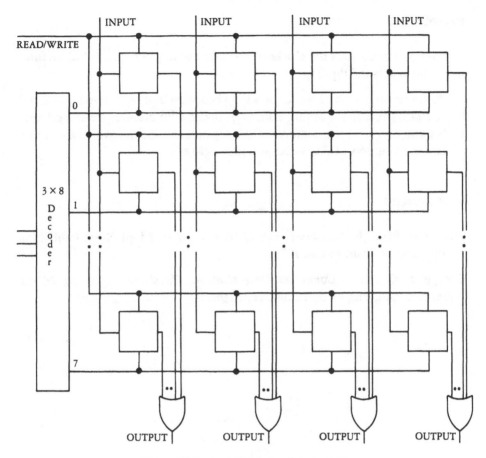

Figure 38.7 A small but complete memory

one of eight memory words. As soon as an address numbered 0 through 7 is sent to the decoder, the word with that address is selected by the mechanism already discussed: the SELECT line on each cell in that word carries a 1.

Given that a particular word has been selected, the presence of a 0 or 1 on the READ/WRITE line determines whether the computer will write a 4-bit number in the selected row or will read the number already contained there. The outputs of all memory cells in each column are tied together into single OR gates, and these feed, in turn, into a special register, treated more fully in Chapter 48, which describes a simple computer.

Problems

1. Design circuitry for the clocked *RS* flip-flop to implement the CLEAR function. This resets the flip-flop.

2. Representing a 4-bit register by a long box with four input and four output lines, attach these to the appropriate lines in the 4-bit memory described here so that one may store the contents of this register in a specific word of memory or load the contents of that word into the register.

References

Charles H. Roth, Jr. *Fundamentals of Logic Design.* 2d ed. West Publishing Company, St. Paul, Minn., 1979.

George W. Gorsline. *Computer Organization: Hardware/Software,* 2d ed. Prentice-Hall, Englewood Cliffs, N.J., 1986.

NONCOMPUTABLE FUNCTIONS

The Busy Beaver Problem

B eavers, well known for their ability to stay with a task until it has been completed, busily ply the water between the forest and the dam, carrying a new stick or branch for the structure with each trip.

Turing machines (see Chapter 31), running back and forth along their tapes, reading a symbol here and writing a symbol there, remind us of beavers. Just how busy can a Turing machine be? Some Turing machines are infinitely busy in the sense that they never halt. Moreover, many of those that do halt may be made busy for arbitrarily long periods by altering the initial appearance of their tape before each run. Thus it seems sensible to frame this question in the context of an intially empty tape, for all machines that halt with such a tape as input.

In 1962, Tibor Rado, a Hungarian mathematician, invented what is now referred to as the *busy beaver problem* (Figure 39.1): Given an *n*-state Turing

Figure 39.1 A busy beaver writes another 1.

machine with a two-symbol alphabet {0, 1}, what is the maximum number of 1s the machine may print on an initially blank (0-filled) tape before halting? There can be no question that this number, which we denote $\Sigma(n)$, exists, for the number of n-state Turing machines is finite.

What makes a problem worth solving, among other things, is its difficulty. The busy beaver problem cannot be solved in general by a computer since the function $\Sigma(n)$ grows faster than any computable function $f(n)$. To see that this is true, suppose B is a Turing machine that computes the busy beaver function $\Sigma(n)$, and suppose that B has q states. This means that when B is confronted with an input tape with the integer n written on it, B outputs the number $\Sigma(n)$. Without loss of generality, we may assume that both n and $\Sigma(n)$ are written in binary notation.

Now let B_n be a Turing machine that has n states and prints $\Sigma(n)$ ones before halting, let C be a machine that converts a number in binary to a number in unary notation, and let A be a machine that converts a blank tape to one with the number n written in binary upon it. The three machines, put together, may be represented by ABC. This is a single machine that starts with a blank tape and halts with $\Sigma(n)$ ones on the tape. Moreover, it has only $\lceil \log n \rceil + q + r$ states, since $\lceil \log n \rceil$ states are required by A and a constant number of states, q and r, are required by B and C, respectively.

For any integer n such that $n > \lceil \log n \rceil + q + r$, machine ABC prints just as many 1s as B_n and yet uses fewer states! However, since it is easy to show that $\Sigma(n) > \Sigma(m)$ when $n > m$, an obvious contradiction is obtained. Since machines A and C clearly exist, B cannot exist. Accordingly, $\Sigma(n)$ is not a computable function.

The table below summarizes the present state of our knowledge about the values of $\Sigma(n)$.

n	$\Sigma(n)$
1	1
2	4
3	6
4	13
5	≥ 4098

The jump from $\Sigma = 13$ at $n = 4$ to $\Sigma \geq 4098$ at $n = 5$ is symptomatic of the uncomputable nature of Σ'. The number 4098 has an interesting story behind it.

The August 1984 issue of *Scientific American* magazine carried an article describing what was then the busiest 5-state beaver known, found by Uwe Schult, a German computer scientist, in 1984. Schult's busy beaver produced 501 ones before halting. In response to the article, George Uhing, an American programmer, conducted a computer search for 5-state busy beavers and found one that printed 1915 ones before halting. Later in 1989, Uhing's record beaver was supplanted by a new, busier beaver discovered by Jürgen Buntrock and Heiner Marxen of Germany. The Buntrock–Marxen beaver, discovered by a three-day search on a high-speed computer, writes 4098 ones before halting! Here is the Turing machine they discovered, written out as a transition table. The notation A0 → B1L means, "If in state A the machine reads a 0, enter state B, write a 1 and move one cell to the left."

$$
\begin{array}{ll}
A0 \rightarrow B1L & A1 \rightarrow A1L \\
B0 \rightarrow C1R & B1 \rightarrow B1R \\
C0 \rightarrow A1L & C1 \rightarrow D1R \\
D0 \rightarrow A1L & D1 \rightarrow E1R \\
E0 \rightarrow H1R & E1 \rightarrow C0R
\end{array}
$$

In general, one could select an arbitrary computable function such as 2^n and write

$$\Sigma(n) \geq 2^n$$

(for large enough n) with every confidence of being right. But there is more.

The theorem which we have just proved was discovered by Rado in 1962. It was soon to be superceded, in a sense, by a rather more dramatic result found by M. W. Green in 1964:

For any computable function f,

$$f(\Sigma(n)) < \Sigma(n + 1)$$

for infinitely many values of n.

Thus, for example, if we take f to be the function

$$f(m) = m^{m^{m^m}}$$

for any fixed number of exponentiations, we nevertheless must conclude that

$f(\Sigma(n)) < \Sigma(n+1)$. In other words,

$$\Sigma(n+1) > \Sigma(n)^{\Sigma(n)^{\Sigma(n)^{\cdot^{\cdot^{\Sigma(n)}}}}}$$

for infinitely many values of n.

It would seem that the most fruitful area in which to attempt solving the busy beaver problem is to improve the lower bounds appearing in the table above. But, as computer searches conducted during the last ten years have shown, this is not any easy task, even when $n = 5$.

One reason for the enormous difficulty of the busy beaver problem lies in the ability of relatively small Turing machines to encode profound unproved mathematical conjectures such as Fermat's last "theorem" or the Goldbach conjecture (every even number is the sum of two primes). Knowledge of whether such machines halt is tantamount to proving or disproving such conjectures. If $\Sigma(n)$ were known for the value of n corresponding to such a conjecture, we would (in principle, at least) be able to decide the conjecture by knowing how long to wait for its corresponding machine to halt.

Some theorists doubt that we shall ever compute even $\Sigma(6)$.

Problems

1. Find the busy beavers having one, two, and three states.

2. Find a not-so-busy beaver, a general prescription for a Turing machine having n states that produces 2^n ones before halting, $n > 4$.

References

Devek Wood. *Theory of Computation.* Harper and Row, New York, 1987.

Fred Hennie. *Introduction to Computability.* Addison-Wesley, Reading, Mass., 1977.

40

HEAPS AND MERGES

The Fastest Sorts of Sorts

It would be unfair to have shown the reader that no algorithm can sort n numbers in fewer than $n \log n$ steps (see Chapter 35) without displaying an algorithm or two that can actually do it that quickly. The most dramatic examples of such algorithms are the techniques known as *heapsort* and *mergesort*. They demonstrate once again the amazing efficiencies that sometimes result from a divide-and-conquer approach to a problem.

Heapsorting depends on a special structure known as a *heap*. A natural example of a heap might be the organizational chart of a corporation in which each employee has talents and abilities summarized by a single number. Far from resembling a heap in the ordinary sense, such a structure would resemble a tree (Figure 40.1).

In a stable organization, each employee's position is reflected by his or her number; it is greater than those supervised and less than the supervisor's. Such a tree is a heap.

The heapsort algorithm depends critically on our ability to convert an arbitrary tree of numbers to a heap. Moreover, the faster we can do this, the faster the algorithm will run. Suppose, then, that we are faced with a corporate hierarchy

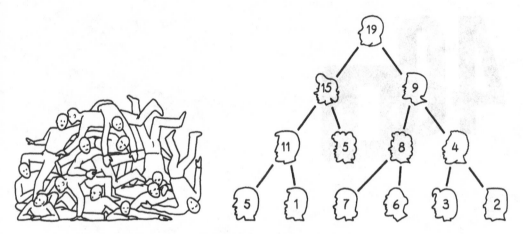

Figure 40.1 A "heap" and a heap

in which a number of employees obviously require promotion. One way to deal with the problem is to continue selecting at random an employee having a higher number than her or his boss and promoting that employee, that is, exchanging the employee with the boss in the corporate hierarchy. Eventually a heap will emerge, but how eventually? For some examples involving *n* employees, one might carry out far more promotions than there are employees!

The most efficient procedure for converting an arbitrary tree to a heap is very simply described in algorithmic terms:

for each node *X*
   W ← max {Y, Z}
   if *X* < *W* **then** exchange *X*
       with the larger of
       Y and *Z*

The recipe is not quite complete: no indication of what order to process the tree's *n* nodes was given. Here, order is crucial. If the algorithm works from the bottom up, however, only 2*n* promotions are needed in the worst case. At the bottom of the tree are a number of small subtrees consisting of exactly three nodes each. Apply the promotion procedure above to each of these. At the next level up, the subtrees have seven nodes each. Apply the promotion procedure to the top three nodes in each of these trees. In general, the promotion is carried out in the setting shown in Figure 40.2.

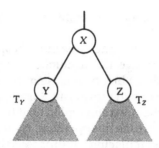

Figure 40.2   Will T_y and T_z be heaps after x is demoted?

If X is already greater than Y and Z, no promotion occurs. Since the subtrees T_Y and T_Z are already heaps, so is the subtree at X. If X is exchanged with Y, however, then T_Z continues to be a heap. But the new subtree T_X (with X replacing Y) may not be a heap: X may be smaller than one of its new descendants. In such a case the basic algorithm is invoked again at X. It continues until X finds a final resting place.

It would seem that the algorithm just described might visit various nodes of the tree more than once. How do we know that it will not require on the order of n^2 exchanges to create the final heap? There is a simple proof that at most n promotions suffice. A diagram makes the central idea of the proof clear: Assume that each time a promotion occurs the number replaced is carried by succeeding promotions all the way to the bottom of the tree. In order to count the total number of promotions visually, there is no harm in assuming that each chain of promotions follows a different path through the tree (Figure 40.3). The fact that such an arrangement is obviously always possible means that the total number

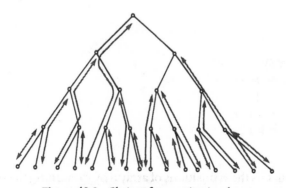

Figure 40.3   Chains of promotion in a heap

of promotions is not greater than the number of edges in the tree. The latter number, of course, is $n - 1$.

Armed with a procedure for making a tree into a heap, we see that heapsorting is straightforward. First, arrange the n numbers arbitrarily in the form of a complete (or nearly complete) binary tree. Next, convert the tree to a heap by means of the algorithm described previously. Sorting is now done by removing the number at the root of the heap and replacing it by a number removed from the bottom of the heap. Each time this operation is carried out, the root number is outputted and the number replacing it is processed by the promotion scheme until it finds its way to an appropriate position in the tree.

The numbers outputted by this version of heapsort will be in decreasing order. The opposite order will be produced if the opposite definition of heap is used—each number is smaller than its descendants.

It takes $O(n)$ steps to convert the initial tree of numbers to a heap. From that point on, for each number removed at the root of the heap, $O(\log n)$ promotions suffice to restore the heap property. The total number of steps required, then, is $O(n \log n)$.

The divide-and-conquer notion enters the heapsort technique at the core of the promotion algorithm: The number at the root of each subtree will be exchanged with at most one of its descendants. In other words, the impact of each promotion will be confined at each stage to just one-half of the descendants (ultimate or not) of a particular node.

The mergesort technique also leans heavily on the divide-and-conquer strategy. If one wishes to sort a sequence A of n numbers in decreasing order, suppose the numbers were already arranged in two equal-size sequences B and C. If both sequences are already sorted in the required manner, how quickly can the original sequence be sorted? The answer is obvious—as long as it takes to merge the two sequences into one. The actual merging algorithm must take relative sizes into account:

```
j ← 1; k ← 1
for i = 1 to n
    if B(j) > C(k)
        then A(i) ← B(j);
             j ← j + 1
        else A(i) ← C(k);
             k ← k + 1
```

The algorithm fills the n positions of array A by examining the relative sizes of the next numbers to be drawn from the B and C arrays. If the B number is larger, choose it. Otherwise, choose the C number. Obviously the algorithm absorbs $O(n)$ steps, $6n + 2$ to be exact (see Chapter 15).

Merging two previously sorted sequences is a long way from sorting a completely unsorted sequence. What is the point of merging? One may answer this question, implicitly invoking the divide-and-conquer concept, by asking another: How did the previously sorted sequences get sorted? The answer comes at once: as the result of merging two sorted sequences of half the length.

Continuing this sequence of questions to the bitter end results in the realization that it can only continue for $\lceil \log n \rceil$ stages. At the first stage, two sequences of length $n/2$ are merged in pairs. Indeed, at each stage n numbers participate (each once) in the merging operation. Clearly, the total number of basic steps taken by the implicit algorithm is $O(n \log n)$.

The algorithm is called mergesort for obvious reasons. It lends itself especially well to a recursive formulation:

MERGESORT

$sort(i, j)$: **if** $i = j$ **then** *return*
           else
               $k \leftarrow \lfloor (i + j)/2 \rfloor$
               $sort(i, k)$
               $sort(k + 1, j)$
               merge $A(i, k)$ and $A(k + 1, j)$
$sort(1, n)$

The algorithm assumes the existence of a merge routine like the one described earlier. It merges the A entries from the ith to the kth positions with those from the $(k + 1)$th to the jth. Recursion enters the picture when the sort procedure is defined in terms of itself. It says, in effect, the following:

> To sort the numbers in the ith to jth positions, find a reasonable approximation k to the index midway between i and j. Then sort the numbers from the ith to kth positions, those from the $(k + 1)$th to the jth, and merge the two sequences (subarrays) just formed.

The final instruction is a single call, $sort(1, n)$ to the procedure with $i = 1$ and $j = n$. This triggers an avalanche of calls originating within the procedure as it invokes itself twice more at every call. In schematic terms, the sequence 2, 8, 5, 3, 9, 1, 6 is sorted as shown in Figure 40.4.

In the divide phase of the mergesort algorithm, the input array A is ruthlessly divided and subdivided until only the individual entries are left, so to speak. In

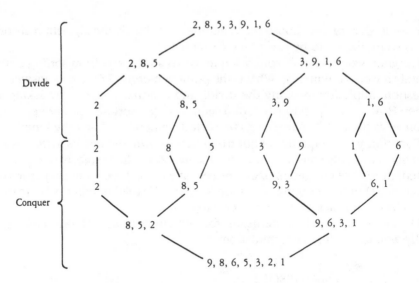

Figure 40.4 Mergesort: the total pattern

the conquer phase, the procedure begins a long sequence of returns from earlier calls to itself. In other words, the innermost sorts are completed, and elements begin to enter the merging process, at first in twos, then in fours, and so on.

Neither the heap structure nor the merge operation appears at first sight to be a key element in a sorting algorithm, yet both are. In both cases, the divide-and-conquer approach leads to a logarithmic factor in sorting time. We simply cannot do better than this with a sequential machine! (See Chapter 62).

Problems

1. Devise a tree of nine numbers in which the maximum number of promotions is effected by the heap-building algorithm. If a manager does not use this algorithm but promotes blindly, what is the greatest number of promotions she might perform in this case? Find formulas for the worst-case time complexity for both methods when there are n numbers in the tree.

2. Write complete heapsort and mergesort algorithms from the descriptions provided. Convert these to programs. Which method turns out to be faster?

References

Thomas A. Standish. *Data Structure Techniques.* Addison-Wesley, Reading, Mass., 1980.

Donald E. Knuth. *The Art of Computer Programming,* vol. III, *Sorting and Searching.* Addison-Wesley, Reading, Mass., 1967.

41

NP-COMPLETENESS

The Wall of Intractability

The term *NP-complete* is often used but not always well understood. In practical terms, an NP-complete problem is one that can be solved by computer only by sometimes waiting an extraordinarily long time for its solution. In theoretical terms, an NP-complete problem is best understood as an application of Cook's theorem. The first NP-complete problem, satisfiability (see Chapter 34), was discovered by Stephen Cook while he was completing his doctorate in computer science at the University of California at Berkeley in 1970. Cook discovered a generic transformation from every problem in a certain large class called NP to a single problem in logic called *satisfiability* (SAT) (Figure 41.1).

In the figure opposite, the action of Cook's generic transformation is illustrated schematically in terms of a few of the better-known problems in NP. The acronym *TSP*, for example, refers to the well-known traveling salesperson problem. Other acronyms such as G3C, VC, PRT stand for problems such as 3-coloring a graph, finding a vertex-covering for a graph, and the partition problem mentioned in Chapter 30. In reality, NP contains an infinite number of problems.

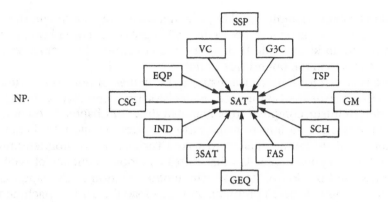

Figure 41.1 Some of the problems that transform to satisfiability

The transformation discovered by Cook is called *generic* because it can be specialized to act on any single problem in NP. For example, it transforms any instance of the traveling salesperson problem (TSP) to an instance of the satisfiability problem in such a way that

1. The transformation can be computed in polynomial time.
2. The instance of TSP has a solution if and only if the corresponding instance of SAT does.

These two properties hold when the generic transformation is specialized to any one of the infinite number of problems in NP: Merely substitute the name of the problem for TSP in 2 above.

The main implication of Cook's theorem is that satisfiability is at least as hard, from a polynomial-time point of view, as any other problem in NP. It must be at least as hard to find a polynomial-time algorithm for satisfiability as for any other problem in NP. From the moment one had a good satisfiability algorithm in hand, one could select an arbitrary problem *ABC* from NP and apply Cook's generic transformation to obtain, in polynomial time, an instance of satisfiability. Applying the satisfiability algorithm would determine the existence of a solution to the satisfiability instance and hence to the instance of *ABC*, by condition 2, all in polynomial time.

Most problems we attempt to solve by computer have solutions that display a structure. The solution to the traveling salesperson problem, for example, would be a minimum-cost route covering all the cities in a given territory. The problems in the class NP have much simpler answers, namely, yes and no. Most problems with a structured answer, like the TSP, can be easily converted to decision problems simply by asking whether a solution of a certain size exists. For example, a decision instance of the TSP might consist of a network of cities and a bound in kilometers. Is there a route covering all the cities with total distance less than the bound, yes or no?

Here N stands for nondeterministic, and P stands for polynomial time. The *class NP* may be defined as the set of all decision problems that can be solved by a nondeterministic computer in polynomial time. In Chapter 26 on nondeterminism, it was shown how to make various conceptual models of computers nondeterministic. It is not difficult to make a Turing machine nondeterministic. Equip it with a guessing facility that writes a random sequence of symbols in some reserved portion of its tape. On another portion of the tape place an instance of some decision problem, and then load the Turing machine with a program to find out, based on the guess, whether the answer is yes or no. If for all yes instances of the problem (and only for these) there is a guess which will cause the Turing machine to answer yes, then the program is said to solve the decision problem. It does so in polynomial time if there is some fixed polynomial that bounds the number of steps in at least one yes computation for each possible yes instance the Turing machine might encounter. In such a case, the problem is said to lie in NP.

As is usual in time-complexity analysis (see Chapter 15), the size of the problem instance serves as the independent variable in the polynomial. Normally in the theory of NP-completeness the size of an instance is the length of symbolic string that encodes it on the Turing machine's tape. One assumes that the representation is reasonably economical since too many unnecessary symbols could distort the true complexity of a computation.

The partition problem (PRT) (see Chapter 30) is readily restricted to decision form: Given a set of positive integers, is there a partition of the set into two parts so that both sum to the same number, yes or no? Figure 41.2 shows how a particular instance of the problem might be solved by a nondeterministic Turing machine.

Here the machine's tape has been divided into three regions, one for the

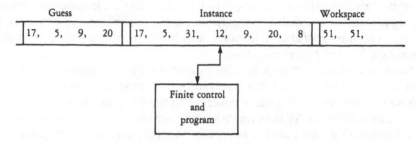

Figure 41.2 A nondeterministic Turing machine solves the partition problem.

random guess, one for the instance, and one for working out intermediate results. Is there a partition of integers 17, 5, 31, 12, 9, 20, 8? Yes there is, and by some magic our nondeterministic Turing machine has specified one of the parts. The sum of 17, 5, 9, and 20 is 51, and the sum of its complement, 31, 12, and 8, is also 51. After making its guess, the Turing machine next operates in deterministic fashion. Moving back and forth across the tape, the machine (directed by its program, of course) first ensures that the integers appearing in the guess also appear in the instance. It then adds the integers in the guess and writes the sum in its workspace. Finally, it adds all the integers in the instance, divides by 2, and then compares the result with the sum written in the workspace. If the numbers are equal, it outputs a yes; otherwise, it outputs a no. Some guesses may cause the Turing machine never to halt, but these (along with the no computations) are not considered in nondeterministic time complexity.

For each yes instance of PRT, there is a yes computation of minimum length. Over all yes instances of size n, one of these minimum length computations is a maximum. This is taken to be the complexity of the procedure for input of size n. Clearly, it bounds the length of a yes computation for each yes instance of that size. If, over all values of n, there is a polynomial that bounds these maxima, then PRT is in NP.

It would be a waste of time to write a nondeterministic Turing machine program for PRT and then analyze it to determine the precise value of the program's complexity for each size of input. We are only concerned with whether that complexity is bounded by a polynomial. Suffice it to say that in the case of PRT the procedure outlined above can be realized by a program that never requires more than $O(n^3)$ steps, where n is the size of instance. Thus EQP is in NP.

In fact, PRT, like SAT, is NP-complete. But not all problems in NP are necessarily NP-complete. For example, if we alter PRT slightly by requiring in addition that each number in one part be at least as large as any number in the other part, then we obtain a new problem. The nondeterministic program for this new problem does everything the program for PRT did. It also checks that each number in the guessed part dominates all numbers remaining in the instance. This, too, can be done in a number of steps, that is, $O(n^3)$. Thus the new problem is in NP.

But one hardly needs to resort to a nondeterministic computer to solve the latter problem. It can be solved deterministically in polynomial time. First, sort the given numbers into decreasing order, and then form their sum. Next, scan the numbers from first to last, accumulating a partial sum in the process. If the partial sum ever equals one-half of the total, then output yes. This algorithm can certainly be made to operate in a polynomial number of steps.

The class P consists of all those decision problems which, like the foregoing problem, can be solved in a polynomial time by a deterministic computer.

The famous question "P = NP?" asks whether the two classes of problems are equal. It is not hard to show that P is a subset of NP, so the real question is whether there are any decision problems in NP that do not have a polynomial-time deterministic solution. In spite of the outrageous computing power of a device that always makes the right guess, we seem unable to prove that it makes any difference!

On the other hand, our inability to solve deterministically and efficiently some problems in NP leads us to suspect that the containment is proper. The prime candidates for examples of problems in NP−P are the NP-complete ones. As has been mentioned already SAT is NP-complete. Before we describe other seemingly intractable problems, it would be worthwhile to be more specific about Cook's theorem and just what it means to be NP-complete.

The heart of Cook's theorem is the generic transformation mentioned earlier. As input it takes an instance of problem ABC, say, and a nondeterministic algorithm that solves ABC in polynomial time. The generic transformation generates a number of logical clauses that describe the behavior of Turing machines in general. It also generates clauses that describe the action of a Turing machine under the direction of an arbitrary program with arbitrary input. In this case we use the ABC-solving Turing machine program and a specific instance of ABC. The program generates these clauses in time $O(p^3(n))$, where $p(n)$ is the time complexity of the (nondeterministic) ABC-solving program. Thus in time $O(p^3(n))$, the generic transformation generates a set of logical clauses that describe the ABC-solving program completely in terms of its actions on the given instance. Things are arranged in Cook's theorem so that the resulting set of clauses has a satisfying assignment if and only if the original instance of ABC was a yes instance.

Suppose now that a being from another planet landed at the United Nations building in New York City and handed the citizens of earth a polynomial-time deterministic algorithm for the satisfiability problem. How would we use it? To solve problem ABC, of course.

First, transform the instance to be solved to an equivalent clause system in $O(p^3(n))$ steps, using Cook's generic transformation. Next, apply the alien algorithm which is able to detect yes instances of SAT in time $O(q(m))$ (a polynomial, of course), where m is the size of SAT instance inputted to it. It is an elementary exercise to recognize that the composition of the two procedures would yield a yes if and only if the original ABC instance were a yes instance. Moreover, the composite procedure would run in

$$O(q(p^3(n))) \text{ steps}$$

(see Chapter 45). This may be a large polynomial, but it is still just a polynomial. Obviously, any decision problem in NP can now be solved in polynomial time. This would mean that P = NP and that all the thousands of hours spent here on earth trying to solve some of the problems in NP were superceded by one brilliant mind from outer space!

However, we think that this is an unlikely event, not only because of our previous negative experiences in looking for fast algorithms, but also because a great many of the problems in NP (the very ones we had been having trouble with, for the most part) turn out to be NP-complete.

Recall that what makes SAT NP-complete is the fact that solving any instance of any problem of NP is equivalent to solving some instance of SAT. The same thing turns out to be true of certain other decision problems.

Problems

1. In the vertex-covering (VC) problem, we are given a graph G and an integer k. Is there a vertex cover for G having k or fewer vertices? A subset of G's vertices qualifies as a *cover* if every edge has at least one of its vertices in the subset. Show that the vertex-covering problem is in NP.

2. Given the time complexity claimed for the nondeterministic algorithm for the partition problem, how big an instance of satisfiability will be produced by Cook's generic transformation? How quickly could we solve PRT if a Martian presented us with an $O(n \log n)$ time algorithm for SAT?

References

M. R. Garey and D. S. Johnson. *Computers and Intractability: A Guide to the Theory of NP-completeness.* Freeman, San Francisco, 1979.

Harry R. Lewis and Christos H. Papadimitriou. *Elements of the Theory of Computation.* Prentice-Hall, Englewood Cliffs, N.J., 1981.

42

NUMBER SYSTEMS FOR COMPUTING

Chinese Arithmetic

We are so used to the decimal notation for numerical calculations we perform in our heads (or hand calculators) that we experience a certain shock when another system is suggested. Think, for example, what might have happened if humans had evolved with four fingers (a thumb and three fingers) on each hand (Figure 42.1).

If such were the case, we might well be using the octal system in which 8 instead of 10 digits would participate. For example, one could count as follows: 1, 2, 3, 4, 5, 6, 7, 10, 11, 12, 13, 14, 15, 16, 17, 20, 21, . . . and so on. Obviously, 10 in the octal system (10_8) would represent the number we write as 8 in the decimal system (8_{10}). Thus 20_8 and 21_8 would be 16_{10} and 17_{10}, respectively.

A great advantage of the octal system lies in 8 being a power of 2. Each octal digit can be represented by a triple of binary numbers in accordance with the following scheme:

$$0_8 = 000 \quad\quad 4_8 = 100$$
$$1_8 = 001 \quad\quad 5_8 = 101$$
$$2_8 = 010 \quad\quad 6_8 = 110$$
$$3_8 = 011 \quad\quad 7_8 = 111$$

It is therefore extremely easy to convert back and forth between octal and binary notation. In going from octal to binary, merely replace each octal digit by the appropriate binary triple. In the reverse direction, count off triples from the right-hand end of the binary numbers, converting each triple to the corresponding octal digit.

$$31{,}572 \leftrightarrow 11\ 001\ 101\ 111\ 010$$

A number X written in the notation of an arbitrary base b consists of digits X_1, $X_2, \ldots, X_n$, where

$$X = X_n b^n + X_{n-1} b^{n-1} + \cdots + X_2 b^2 + X_1 b + X_0$$

This scheme embraces not only binary and octal numbers but virtually all positional notations.

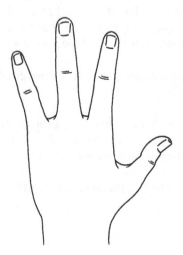

Figure 42.1 A hand suited to the octal system

Binary numbers form the basis of computer arithmetic because the elements of memory and logic all have two states that can be identified as 0 and 1 (see Chapter 13). This does not preclude other arrangements, however. For example, three-state components are almost as easy to build. They open the way for ternary (base-3) arithmetic or even a special system known as *balanced ternary*.

In balanced ternary, the symbols are $\bar{1}$, 0, and 1. These symbols play a role similar to, but not identical with, 0, 1, and 2, respectively. They are sometimes called *trits* in analogy to *bits*. Addition is defined in the following table.

$$
\begin{array}{ccccccccc}
\bar{1} & \bar{1} & \bar{1} & 0 & 0 & 0 & 1 & 1 & 1 \\
+\ \bar{1} & 0 & 1 & \bar{1} & 0 & 1 & \bar{1} & 0 & 1 \\
\hline
\bar{1}\bar{1} & \bar{1} & 0 & \bar{1} & 0 & 1 & 0 & 1 & 11
\end{array}
$$

In two cases above, an extra trit is generated by addition. In serial addition this would imply a carry to the next sum. By adding 1, we may start from 0 and generate the positive balanced ternary numbers in succession; or we may subtract 1 and generate negative numbers:

	0	1	2	3	4	5	6	7	8	9	10	11	
Positive	0	1	$1\bar{1}$	10	11	$1\bar{1}\bar{1}$	$1\bar{1}0$	$1\bar{1}1$	$10\bar{1}$	100	101	$11\bar{1}$	$\cdots$
Negative	0	$\bar{1}$	$\bar{1}1$	$\bar{1}0$	$\bar{1}\bar{1}$	$\bar{1}11$	$\bar{1}10$	$\bar{1}1\bar{1}$	$\bar{1}01$	$\bar{1}00$	$\bar{1}0\bar{1}$	$\bar{1}\bar{1}1$	$\cdots$

This version of ternary numbers is called balanced because of the symmetry between 1 and $\bar{1}$: The negative of a given balanced ternary number is obtained by interchanging the two trits, 1 and $\bar{1}$, throughout the number. Moreover, the sign of a number is positive if 1 is the leading trit and negative if $\bar{1}$ is the leading trit.

A given number in ternary notation is converted to balanced ternary very simply:

1. Add 1 to each digit from right to left, propagating carrys and even adding 1 to the last carry, should it be propagated past the leading digit.
2. Subtract 1 from each digit without carrys.

By this recipe the number 211020011 becomes:

1. 2022201122
2. $1\bar{1}1\bar{1}1\bar{1}0011$

Another numerical representation with some promise for fast computer arithmetic involves a set of positive numbers called moduli $m_1, m_2, \ldots, m_n$

that are relatively prime in pairs. Given an integer x that is not too large (less than the product m of the moduli), it may be represented uniquely by its modular digits:

$$x_i \equiv x \bmod m_i \qquad i = 1, 2, \ldots, n$$

If two numbers x and y have the same such representations, then $x \bmod m_i = y \bmod m_i$ for each i. In other words, $x - y \equiv 0 \bmod m_i$ for each i. It follows that $x - y \equiv 0 \bmod m$. Since both x and y are less than m, they must be equal.

It is also true that any set of numbers x_i such that $x_i < m_i$ represents an integer in the modular system. These two observations comprise the famous *Chinese remainder theorem,* the major theoretical underpinning for modular or Chinese arithmetic. The second observation, however, is not as easy to prove as the first, especially if we seek a proof that leads to a fast and convenient computation of a number from its modular representation.

The fact that any modular string $x_1 x_2 \cdots x_n$ can be converted to a decimal number is best demonstrated by a fast algorithm that actually carries out the conversion. The first step in the algorithm is to find a set of constants a_{ij} such that for each pair of moduli m_i and m_j we have

$$a_{ij} \times m_i \equiv 1 \bmod m_j$$

In more general terms, this amounts to finding two integers a and b such that

$$a m_i + b m_j = 1$$

Since 1 is the greatest common divisor of m_i and m_j, the coefficients a and b are readily computed by using Euclid's algorithm (see Chapter 10). In this case, we may take a as a_{ij}, so that

$$a_{ij} \times m_1 = 1 - b \times m_j$$

Armed with the constants a_{ij}, one for each pair of indices i, j such that $i < j$, we can compute the number in two stages by the following double-loop algorithms:

$y_1 \leftarrow x_1$
for $j \leftarrow 2$ **to** n
    $y_j \leftarrow x_j - y_1$
    for $i \leftarrow 1$ **to** $j - 2$
        $y_j \leftarrow y_j \times d_{ij} - y_{i+1}$
    $y_j \leftarrow y_j \times a_{j-1,j} \bmod m_j$

When the numbers y_j have all been computed, a second stage computes x:

$$x \leftarrow y_1$$
$$\textbf{for } j \leftarrow 2 \textbf{ to } n$$
$$\quad z \leftarrow y_j$$
$$\quad\quad \textbf{for } i \leftarrow 1 \textbf{ to } j-1$$
$$\quad\quad\quad z \leftarrow zm_i$$
$$x \leftarrow z + x$$

The foregoing algorithm constitutes a proof of the Chinese remainder theorem, but only if one also proves that the number x computed in the second stage has the appropriate residues.

The operations of addition, subtraction, and multiplication are especially easy to perform in modular arithmetic. In terms of individual residues x_i and y_i, the ith residues of the sum, difference, and product of two numbers x and y are, respectively

$$(x_i + y_i) \bmod m_i$$

$$(x_i - y_i) \bmod m_i$$

$$(x_i \times y_i) \bmod m_i$$

In each case the operation can be carried out on each pair of residues independently of the others. This suggests that an especially fast arithmetic unit could be built to calculate the n residues of the sum (or difference, or product) in parallel.

Against such startling efficiencies, however, one must weigh certain disadvantages. For example, given two modular numbers, which is larger? Is there an algorithm or piece of hardware that can decide this faster than by converting the two numbers to decimal form and comparing by conventional means? These and other problems have undoubtedly prevented modular arithmetic from being implemented on a large scale in modern computers.

Problems

1. Derive subtraction tables for trits, and find an algorithm that converts a balanced ternary number to ordinary ternary notation.

2. Prove that the number x computed by the two-stage algorithm from x_1, $x_2, \ldots, x_m$ actually has the latter integers as its residues.

3. Modular arithmetic can be extended to include negative numbers by changing the allowed range from $[0, m)$ to $[-m/2, m/2)$. In other words, numbers that range in value from minus one-half the product $m_1 m_2 \cdots m_n$ to its positive counterpart only are permitted. Devise an algorithm that decides whether a number in this range is negative. Naturally, the algorithm has access only to the residues of the numbers.

References

Donald E. Knuth. *The Art of Computer Programming,* vol. 2: *Seminumerical Algorithms.* Addison-Wesley, Reading, Mass., 1969.

Underwood Dudley. *Elementary Number Theory.* W. H. Freeman and Company, San Francisco, 1969.

STORAGE BY HASHING

The Key Is the Address

There are three main techniques for storage and retrieval of records in large files. The simplest involves a sequential search through the n records and requires $O(n)$ steps to retrieve a particular record. If the n records are specially ordered or stored in a tree, then a record can be retrieved in $O(\log n)$ steps (see Chapter 11). Finally, if certain information from each record is used to generate a memory address, then a record can be retrieved in $O(1)$ steps, on average! The last technique, the topic of this chapter, is known as *hashing*. A record in a "typical" file consists of a key and a data item, as in the following example:

3782-A: 670-15 DURALL RADIAL, 87.50, STOCK 24

Here, the key is 3782-A, and the data item consists of a size and make of tire followed by its unit price and the number in stock. In matters of storing, searching, and retrieving records from files, it is enough to specify how one manipulates the key; the data item "tags along," so to speak, given the appropriate programming.

To *hash* a key is to chop it up and use just part of it, in a sense. The part used directly generates a memory address, as in Figure 43.1, which uses two-letter English words as keys.

The last two digits of the ASCII for each key k are used as address $h(k)$ in which each word can be stored. Accordingly, on the extreme right of this figure is shown a portion of computer memory in schematic form. Beside each memory location the keys which hash to that address are shown. Some locations have no keys, and others have more than one.

The latter event is called a *collision,* and to make hashing work properly, collisions must be resolved. One might think that, other things being equal, collisions are relatively rare. Certainly, the large number of collisions in the previous example is due to the fact that several of the two-letter words used as keys end in the same letter. In fact, as Donald E. Knuth points out (see reference at end of this chapter), collisions become almost the rule, even with randomly distributed keys, long before the memory space allocated for hash storage is used up. Knuth illustrates this point with the famous *birthday paradox:* What is the probability that at least two people in a room of 23 persons will have the same birthday? The answer, surprisingly, turns out to be better than .5. If we think of the 365 possible birthdates as memory locations, the 23 people as data items, and their birthdays as keys, then this simple conundrum affords us a penetrating insight into the likelihood of collisions: The probability of at least one collision is more than .5 before even 10 percent of the 365 memory locations are used up!

There are two main methods for resolving collisions, namely, *chaining* and

Key	ASCII	Address
TO	124117	17
IT	111124	24
IS	111123	23
AS	101123	23
AT	101124	24
IF	111106	06
OF	117106	06
AM	101115	15
BE	102105	05
DO	104117	17
AN	101116	16
GO	107117	17
SO	123117	17

Address		Keys
05		BE,
06		IF, OF
07		
10		
11		
⋮		
14		
15		AM
16		AN
17		TO, DO, GO, SO
⋮		
23		AS, IS
24		IT, AT
25		

Figure 43.1 Hashing on digits

289

open addressing. But before describing them, we will see what can be done about keeping the number of collisions to a minimum.

The storage medium is taken to be a set of integer addresses from 0 to M; these can be array indices or actual memory addresses, depending on the level of language used. A key k is generally an alphanumeric string, but we can always convert k to an ASCII-equivalent integer. Assuming that k is already an integer, how do we generate a memory address in the range 0 to M so as to minimize the probability of collisions? Not surprisingly, this question is closely related to that of generating random numbers *ab initio* (see Chapter 8). A reasonably good function h involves division modulo M:

$$h(k) = k \bmod M$$

The only way we can control the function h in this case is to select M; it is not a question of how large or small to make M. It is only necessary that M be the right *kind* of number. It turns out that if M is a prime number, then h yields well-distributed addresses provided that M does not have the form $r^k + a$, where r is the radix of the character set ($r = 128$ for ASCII characters) and a is a small integer. If we take this advice for the miniature example discussed earlier, then $M = 23$ is certainly far away from any number of the form 128^k.

With this value of M we get quite different results for the same 13 two-letter words. Memory addresses in Figure 43.2 are in decimal notation. With the first hash function we suffered five collisions (counting TO, DO, SO, and GO as three collisions), but with the second hash function the number has dropped to just three.

Besides the division method, there is another useful method which uses multiplication. It is based on the general observation that if one takes an irrational number x, forms n multiples of it

$$x, 2x, 3x, \ldots, nx$$

and then takes the fractional parts of these

$$x_1, x_2, x_3, \ldots, x_n$$

the resulting numbers are all different and divide the unit interval into $n + 1$ subintervals. Each of these subintervals has one of at most three possible lengths, and if a new point x_{n+1} is added, it falls into a subinterval of the largest kind. It happens that if x is chosen to be the so-called golden mean (with approximate value 0.61803399), then the three lengths suffer the least variation and are closer together than the lengths produced by any other irrational number.

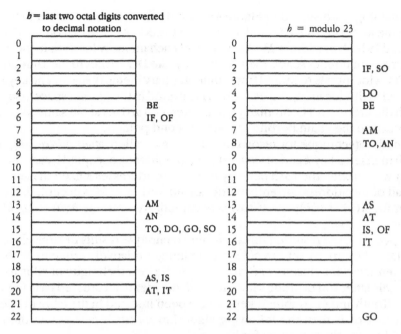

Figure 43.2   Hashing modulo a prime yields an improvement

These observations suggest a hash function h produced according to the following recipe. Let g be the golden ratio, as closely as one's word size may approximate it.

1. $h \leftarrow k \cdot g$
2. $h \leftarrow$ fractional part of h
3. $h \leftarrow h \cdot M$
4. $h \leftarrow$ integer part of h

This calculation forms the product $k \cdot g$, takes its fractional part, scales it up to M, and then takes the nearest integer to (and less than) the result.

If we apply this method to our ongoing example of two-letter keys, just two collisions are obtained, which is about as good a performance as we could expect under the circumstances.

Having discussed two kinds of hash functions, we turn now to two techniques for handling collisions. The first involves constructing a chain of pointers from each address at which a collision occurs, and the second involves moving to a new address within the hash table that portion of memory devoted to the storage of keys.

Chaining is illustrated in Figure 43.3 for the hash function in our first example; the keys TO, DO, GO, and SO all hash to address 17 in that order. The keys AS and IS hash to address 23. If a portion of each memory location in the table is reserved for pointer space, then when a key like DO hashes to the same address as TO, a location is recruited from some auxiliary memory space. The key DO is placed in this location, and its address is inserted in the pointer space associated with the first word TO. Similarly when, later, GO arrives at the same location, it is added to the chain beyond DO via a second pointer.

Another technique for resolving collisions is called *open addressing*. If we wish to insert a key k and discover that $h(k)$ is already occupied by another key, then we "probe" the hash table by examining addresses a fixed distance $p(k)$ ahead of the address currently being considered. If $h(k)$ is occupied, then we examine $h(k) - p(k)$; and if *that* is occupied, we examine $h(k) - 2p(k)$, and so on.

If $p(k)$ is 1, the resulting linear probing technique results in large clusters as the table fills up. By setting $p(k) = c$, an integer relatively prime to M, there is less tendency for clusters to form. Perhaps the best technique of all is to combine this latter sort of probe sequence with the idea of an ordered hash table due to D. Knuth and O. Amble: If keys k have been inserted in the table in decreasing order of k, then the following algorithm inserts a new key in the table, provided that there is room for it.

1. $i \leftarrow h(k)$
2. **if** *contents*$(i) = 0$
           then *contents*$(i) \leftarrow k$; exit
3. **if** *contents*$(i) < k$
           then swap values of *contents*(i) and k
4. $i \leftarrow i - p(k) \bmod M$
5. **go to** 2

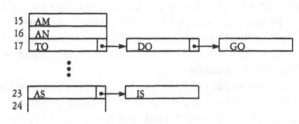

Figure 43.3

Interestingly enough, if this insertion policy had been followed from the very beginning of the table's construction, it would have produced precisely the same table we are assuming! In other words, this algorithm both introduces and maintains the same table we would obtain by inserting first the largest key, then the next largest key, and so on. In the words of Thomas A. Standish (see the reference at the end of this chapter),

This process more or less resembles a game of "musical chairs" in which an unseated child is trying to find an empty chair to sit in while the rest of the children are already seated. The unseated child examines chairs in some order (his personal probe sequence, let us say) until locating an empty chair to sit in, or until finding a smaller child. If he finds a smaller child he bullies him into releasing his seat, and sets the smaller child roaming on his own (probably different) personal probe sequence looking for an empty seat or a smaller child to bully into an exchange.

Of course, one does more with a hash table than merely make insertions. More frequently one is searching a table to see if a given key is there. Occasionally one is also deleting items from a table.

For the most part, searching a hash table follows the same route as inserting a new key: Form the hash value $h(k)$, and check that address to see if k is in the table. Depending on whether chaining or open addressing has been used, one next follows a chain of pointers or a probe sequence within the table.

If m of the M addresses in a table contain keys and the chains in a table contain a further n keys, then the number of steps required to search a chained table is n/m. If the hash function leads to a relatively uniform distribution of keys, then m will be larger than n and the search time will be essentially constant (and small).

The ratio m/M is called the *load factor* in a hash table. Under the open-addressing technique, a linear probe sequence will require

$$O\left(\frac{M}{M-m}\right) \text{ steps}$$

to locate a key already in the table. Although this result can be derived theoretically, so far there is only empirical evidence to suggest that an ordered hash table requires

$$O\left(\frac{M}{m}\log\frac{M}{M-m}\right) \text{ steps}$$

for a successful search. This average complexity bound is, nevertheless, conjectured. Note that the linear probe complexity grows much faster than the conjectured ordered hashing complexity as m gets close to M.

It is useful to compare various storage and retrieval techniques in coming to a decision about which is best for a particular application. As an example, we may compare hashing with binary tree methods (see Chapter 11) and conclude that although the access time is much smaller in hashing, there is generally a fixed table size M. This inflexibility may be a disadvantage if the file to be maintained has no specific growth limit. However, hashing is easier to program than some binary tree-based schemes.

Problems

1. Prove that the probability that at least 2 of 23 randomly selected people will have a common birth date is greater than .5. What is the probability of at least one collision among 13 random keys in a hash table with 23 addresses?

2. Compute the hash values of the 13 two-letter keys, using the multiplicative method. Devise a computer experiment to compare this method with the division method.

3. Devise an algorithm to search an ordered hash table for a key K. What happens when the probe sequence reveals a key smaller than the one sought?

References

D. E. Knuth. *The Art of Computer Programming*, vol. 3: *Searching and Sorting*. Addison-Wesley, Reading, Mass., 1973.

T. A. Standish. *Data Structures and Techniques*. Addison-Wesley, Reading, Mass., 1980.

CELLULAR AUTOMATA

The Game of Life

In a darkened room, eager faces peer at a steadily evolving pattern of little white squares on a display screen. Within this pattern, some populations of squares may be growing while others appear headed for extinction; thus the name *life* given by its creator, John Conway, a Cambridge mathematician, to a game that has intrigued millions. When it was first described by Martin Gardner in the October 1970 issue of *Scientific American,* the game quickly established itself as a major spare-time preoccupation of senior and graduate students having access to a graphics computer. Even some faculty found themselves drawn into the magic circle surrounding those scintillating screens, perhaps watching something like the succession of patterns in Figure 44.1.

The game of Life is an example of a cellular automaton. Formally, we must think of an infinite square grid in which each cell exists in one of two states, "living" or "dead." Each cell is a simple automaton that at every tick of a great clock must decide which state it will be in until the next tick. It makes this decision on the basis of not only its present state but also those of its eight neighbors, four adjacent along sides and four adjacent at corners. Here are the rules on which that decision is based:

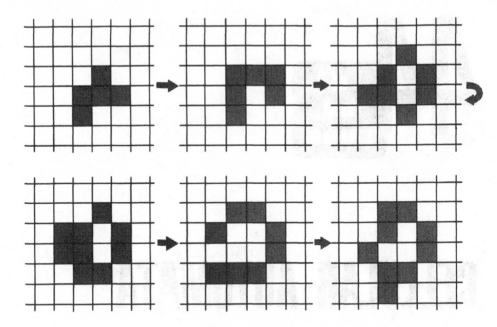

Figure 44.1 Six generations of Life

1. If the cell is alive at time t, it will remain alive at time $t + 1$ if it is not "overcrowded" or "undernourished." In other words, it must have at least two living neighbors and no more than three.

2. If the cell is dead at time t, it will remain dead unless it has three "parents." That is, the cell must have three living neighbors in order to be born again.

Although they sound almost arbitrary to the uninitiated, Conway went to some trouble to discover rules which made the behavior of populations of live cells as difficult as possible to predict. On the way, he experimented with dozens of different rules, discarding each set in turn. Among the notes he sent to Gardner for the 1970 article was the prediction that no population could grow without limit — sooner or later, every population either would become extinct (all cells eternally dead) or would fall into an endlessly repeating cycle of patterns.

The game of Life can be played by hand on a ruled grid using counters. It is difficult, however, to play the game with just one color of counter because one must remember, in going from one generation to the next, just which counters were present in the former. For this reason, it is much better to use two colors, one for already living cells (say, white) and the other for newly born cells (say, gray). Given the t generation of white counters, first go around putting in gray counters wherever a new cell is born. Then remove white counters representing cells which are to die. Finally, replace the gray counters by white ones.

Of course, if one has access to a computer, much more can be done. Programming the game of Life is reasonably straightforward, and even if the computer has no graphics terminal attached, successive generations can be printed as patterns of X's on a printer. In some ways, having hard copy is an advantage, especially for examining past populations for the purpose of detailed analysis. It was by using a computer that Conway and many others caught up in the excitement were able to find novel and interesting patterns. An initial pattern of live cells would be typed into the computer and appear on the screen, and with a press of a key, successive generations of the pattern would appear.

Figure 44.2 shows *(a)* the evolution (with the 13 intermediate generations not shown) of a row of seven live cells into a "honey farm" consisting of four "beehives" and *(b)* a five-cell configuration called a *glider*. The glider repeats itself every 4 cycles but ends up in a new position!

With so many people playing Life, it was not long before someone discovered a counterexample to Conway's conjecture that no populations could grow without limit. A group of six students at the Massachusetts Institute of Technology discovered a "glider gun," a configuration that emits a glider every 30 generations. A pattern which grows into the glider gun after 39 moves is shown in Figure 44.3. With gliders flying out of the gun every 30 moves, the total number of live cells obviously grows without limit. The same group of students managed to arrange 13 gliders crashing together to form a glider gun!

The algorithm displayed below computes successive generations for Life in matrix L. A 1 in the i, jth entry represents a live cell, and a 0 represents a dead cell.

LIFE

```
1. for i ← 1 to 100
    1. for j ← 1 to 100
        1. s ← 0
        2. for p ← i − 1 to i + 1        /compute effect
             for q ← j − 1 to j + 1      /of neighbors
                s ← s + L(p, q)
        3. s ← s − L(i, j)
        4. if (s = 3) or (s + L(i, j) = 3)
             then X(i, j) = 1            /store life or death
             else X(i, j) = 0            /in auxiliary array X
2. for i ← 1 to 100
    1. for j ← 1 to 100
        1. L(i, j) ← X(i, j)            /refresh L
        2. display L(i, j)              /display L
```

297

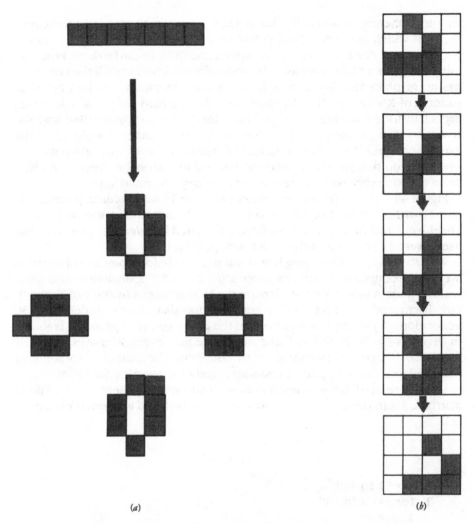

(a)   (b)

Figure 44.2 Evolution of a honey farm (left) and motion of a glider (right)

In writing a program based on this design, the syntax peculiar to the language used must, of course, be substituted for the replacement arrows (representing assignment statements) and indentations (indicating the scope of **for** and **if** statements).

This algorithm is relatively straightforward. The first double-**for** loop computes new values for the (i, j)th entry of L by scanning the 3×3 neighborhood of (i, j). The variable s contains the sum of these values (with the value of $L(i, j)$ subtracted away). The statement 1.1.4 embodies the life-and-death criteria mentioned earlier. The auxiliary array X holds the new value of $L(i, j)$. This value cannot yet be placed in L since the old value of $L(i, j)$ must participate in four more computations of L values. Only at step 2.1.1 do the new L values (carried in X) finally replace the old ones. At the same time, so to speak , the new values are displayed on one's screen at line 2.1.2.

This algorithm computes just one generation of Life. It must be embedded in yet another loop that the programmer may structure according to taste: a fixed number of generations may be computed or the program may be terminated by a keystroke.

Living configurations that reach the edge of the 100 $\times$ 100 matrix boundary will automatically die. One may prevent this by "wrapping the matrices around" using modular arithmetic. For example $L(100,1)$ will be adjacent to $L(1,1)$. In this case, however, the computation of s in the loop at 1.1.2 must be modified to reflect the new rules of cell adjacency.

Any program based on this algorithm must include the appropriate initialization statements.

The academic area called cellular automata has fascinated many computer scientists and mathematicians since John von Neumann invented the subject in 1950. His aim was to construct a self-reproducing "machine." On being persuaded that cellular spaces were an ideal setting for gedanken experiments in this area, von Neumann arrived, finally, at a cellular automaton each cell of which had 29 states. He proved the existence of (but did not explicitly state!) a configuration of about 200,000 cells which would self-reproduce. This meant that when the cellular automaton was put into this configuration, it would, after a definite period, result in two such configurations, side by side. Since von Neumann's time, much simpler self-reproducing cellular automata have been found.

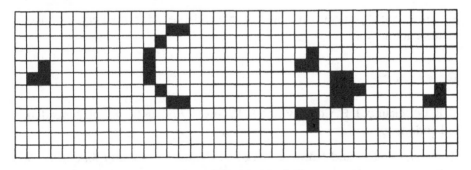

Figure 44.3 A glider gun

In formal terms, a *cellular automaton* consists of three things:

1. An automaton, a copy of which is associated with each cell in an infinite *n*-dimensional grid.
2. A neighborhood function that specifies which of the cells adjacent to a given one may affect it.
3. A transition function that specifies, for each combination of states in "neighboring" cells, what the next state of the given cell will be.

Besides constructing self-reproducing automata, researchers have developed cellular automata which can carry out computations. This is usually done by embedding the equivalent of a Turing machine in a cellular space: Moving patterns of states represent the read/write head, scanning what other state patterns represent as a tape. There are, at least in "mental" existence, universal computational cellular automata, even ones that self-reproduce (see Chapter 46). Indeed, it was recently shown that the game of Life itself has this property!

Problems

1. Write the Life algorithm as an actual program in a language of choice.

2. A one-dimensional cellular automaton coinsists of an infinite strip of cells. A great advantage of such automata is that one can watch the history of a given binary configuration unfold on a display screen if successive generations are displayed as successive rows. Write a program to do exactly this. Use neighborhoods consisting of two cells on either side of the central one. Employ the following rule: If two or four of the cells in a given cell's neighborhood are alive (state 1) at time t, the given cell will be alive at time $t + 1$. Otherwise, it will be dead. Note that each cell is a member of its own neighborhood.

3. How would you convert a one- or two-variable differential equation into a cellular automaton?

References

E. R. Berlekamp, J. H. Conway, and R. K. Guy. *Winning Ways,* vol. 2, Academic, London, 1982.

Tommaso Toffoli and Norman Margolis. *Cellular Automata Machines.* MIT Press, Cambridge, MA, 1987.

COOK'S THEOREM

Nuts and Bolts

ook's theorem establishes that the satisfiability problem (see Chapter 34) is NP-complete. It does this by displaying a generic transformation that in polynomial time maps each and every problem in NP (see Chapter 41) into the satisfiability problem. The details of that transformation, its nuts and bolts, consist merely of several systems of clauses that express in logical form the operation of a Turing machine. For each problem in NP, the Turing machine involved is the nondeterministic one that solves the problem in polynomial time. In any event, the generic transformation can be viewed largely as the expression in logical language of everything that a Turing machine can (and cannot) do (Figure 45.1).

The stress here is on the word *everything*. Things you would hardly regard as worth mentioning must be expressed in the clause systems to be constructed. For example, a Turing machine of the type used in Cook's theorem (and elsewhere in this book) can be in only one state at a time, not two.

We use predicate notation (see Chapters 5 and 58) to express the current state of affairs in a Turing machine. The predicate $S(t, q)$ will have the value true (or 1) if at time t our Turing machine happens to be in state q. If not, it has the value

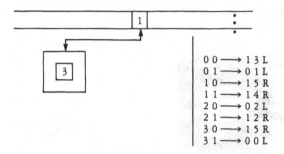

$$
\begin{array}{ll}
0\ 0 \longrightarrow 1\ 3\ \mathrm{L} \\
0\ 1 \longrightarrow 0\ 1\ \mathrm{L} \\
1\ 0 \longrightarrow 1\ 5\ \mathrm{R} \\
1\ 1 \longrightarrow 1\ 4\ \mathrm{R} \\
2\ 0 \longrightarrow 0\ 2\ \mathrm{L} \\
2\ 1 \longrightarrow 1\ 2\ \mathrm{R} \\
3\ 0 \longrightarrow 1\ 5\ \mathrm{R} \\
3\ 1 \longrightarrow 0\ 0\ \mathrm{L}
\end{array}
$$

Figure 45.1   How does one specify a Turing machine's operation in logical terms?

false (or 0). Because we are really trying to capture the operation of a (nondeterministic) Turing machine that requires only $p(n)$ steps for some polynomial p, time t will run from 0 to $p(n)$. The possible values for q, on the other hand, will vary through the set Q of states available to the Turing machine under consideration.

To express the idea that at each time the machine is in only one state, we write the following clause for all possible values of the variables as indicated:

$$[S(t, q) \rightarrow \sim S(t, q')] \quad q, q' \in Q, q \neq q'; t = 0, 1, \ldots, p(n)$$

This clause can be interpreted as follows: If at time t the state is q, then at time t the state cannot be any q' which is not equal to q. One might object that the resulting $(p(n) + 1) \times (|Q|(|Q| - 1))$ clauses are not in the proper form; clauses in an instance of satisfiability must all be in disjunctive form. This situation is easily corrected, however, by using the well-known equivalence of the following two logical expressions:

$$A \rightarrow B \quad \text{and} \quad \sim A \lor B$$

Thus, the general cluase can be rewritten as $[\sim S(t, q) \lor \sim S(t, q')]$.

The final logical expression under construction must also in some way encode the fact that each cell of the tape can hold only one symbol at a time. For this task, we use the predicate $T(t, c, s)$. This predicate is true if at time t cell c contains the symbol s; otherwise, it is false. To express the uniqueness of cell contents at any time, we seem to need a somewhat larger clause system than the one expressing uniqueness of state.

$$[T(t, c, s) \rightarrow \sim T(t, c, s')] \quad t = 0, 1, \ldots, p(n)$$
$$c = -p(n), -p(n) + 1, \ldots, -1, 0, 1, \ldots, p(n)$$
$$s, s' \in \Sigma, s \neq s'$$

Here, Σ denotes the tape alphabet of the Turing machine.

When all the possible values of t, c, and s are substituted into this generic clause, $p(n) \times (2p(n) + 1) \times |\Sigma|(|\Sigma| - 1)$ clauses result. Here again, it is a simple matter to convert the implication to a disjunction. In general, the generic transformation is easiest to understand if the most natural logical operations are used. The clauses appearing here can always be converted efficiently (in polynomial time) to disjunctive form.

It is also an obvious fact of Turing machine operation that at any time the read/write head is scanning one and only one cell. To express this idea, one may use a predicate such as $H(t, c)$. This is true if at time t the Turing machine scans cell number c. Again, the same style of clause can be used with precisely the same kind of effect:

$$[H(t, c) \rightarrow \sim H(t, c')] \quad t = 0, 1, \ldots, p(n)$$
$$c, c' = -p(n), -p(n) + 1, \ldots, p(n); c \neq c'$$

Substituting the values for t, c and c' indicated, we see that the total number of clauses generated is $(p(n) + 1) \times (2p(n) + 1)$. In this system, as in the previous one, note that the cells run from $-p(n)$ to $+p(n)$. This simply means that in the time allotted to the machine, it cannot scan any farther left than the cell at $-p(n)$ or any farther right than cell number $p(n)$.

The foregoing clauses guarantee, in effect, that at no time can more than one thing be happening in the Turing machine. This is the essence of the sequential machine and is also essential to the general guarantee that the machine will do what it is supposed to do. At the same time, one must ensure that the machine does not do things it is not supposed to do — such as changing tape symbols that are not even being scanned. Two of the predicates already introduced serve this function in the following set of clauses:

$$[(T(t, c, s) \wedge \sim H(t, c)) \rightarrow T(t+1, c, s)] \quad t = 0, 1, \ldots, p(n) - 1$$
$$c = -p(n), \ldots, p(n)$$
$$s \in \Sigma$$

At time t cell c may contain symbol s, and yet the read/write head is not scanning cell s. In such a case the symbol remains unchanged in cell c at time $t + 1$. Using the implication formula and DeMorgan's laws (see Chapter 13), one can easily convert this clause to disjunctive form.

At time $t = 0$, the Turing machine, by convention, scans cell 0. We thus assert, in a clause all by itself, this fact:

$$[H(0, 0)]$$

Other initial conditions include the initial state, by convention 0,

$$[S(0, 0)]$$

At time $t = 0$, however, the Turing machine has two important things on its tape. First, a number of cells to the left of cell 0 contain a set of guessed symbols placed there, if one wishes, by some random agency. This idea is expressed by omission: Write no conditions governing the contents of cells to the left of cell 0. We can actually assume that a specific number of these cells contain the guessed symbols and that all the symbols beyond these out to cell number $-p(n)$ contain zero.

Second, cells 1 through n are assumed to contain an instance of the problem at hand:

$$[T(0, i, s_i)] \qquad i = 1, 2, \ldots, n$$

This means simply that at time 0 the ith cell contains s_i, namely, the ith symbol in the instance string inputted to the generic transform.

If the Turing machine has m states, it does no harm to assume that its halting state is the mth. Just as we had to specify the initial conditions of operation, so we must specify what they are like when the time allotted to the computation has elapsed.

$$[S(p(n), m)]$$

The Turing machine must have halted by time $p(n)$. It must also have answered yes. This is symbolized by the appearance of a 1 in cell 0 by that time (a no can be symbolized in the same way by a 0):

$$[T(p(n), 0, 1)]$$

The remaining clause systems make direct or indirect references to the Turing machine program. As mentioned in Chapter 31, the program consists of a collection of quintuples having the following form:

$$(q, s \rightarrow q', s', d)$$

This is interpreted to mean that whenever the Turing machine is in state q and reads the symbol s on the tape, it enters state q', writes an s' in place of the s, and then moves one square in the direction d. This direction is indicated simply by R for right or L for left.

Three separate sets of clauses specify the three possible effects of being in a given state and reading a particular symbol. A new state must be entered, a new symbol written, and a direction taken by the read/write head.

$$[(S(t, q) \wedge T(t, c, s) \wedge H(t, c)) \rightarrow S(t+1, q')] \qquad t = 0, 1, \ldots, p(n)$$
$$[(S(t, q) \wedge T(t, c, s) \wedge H(t, c)) \rightarrow T(t+1, c, s')] \qquad c = -p(n), \ldots, P(n)$$
$$[(S(t, q) \wedge T(t, c, s) \wedge H(t, c)) \rightarrow H(t+1, c')] \qquad \begin{matrix} q \in Q \\ s \in \Sigma \end{matrix}$$

For each combination of the four variables t, c, q, and s, there is a unique value for q', s', and c' specified by the program. It is understood that in each case these are the values to be used in generating the clauses. Indeed, q and s alone uniquely determine these values. The transformation computes the values by merely looking them up in a table containing the nondeterministic program.

This completes the specifications of the clauses generated by the generic transformation in response to a specific program and problem instance. The clauses can be generated in polynomial time in whichever model of computing (see Chapter 66) one adopts. By using a random access machine running a fairly high-level language, a series of nested loops with indices limited in the manner above is perfectly capable of churning out the clauses about as quickly as it can print them.

Correctness of the foregoing transformation is not proved here. Such a proof hinges largely on observing that the clauses limit the virtual Turing machine to behaving just as it should. If this is the case, the clauses can be satisfied only if the machine halts before the polynomial time limit with a 1 occupying cell 0. This, in turn, is possible only if the original problem instance has the answer yes.

The rather large (but still polynomial-sized) set of clauses generated by Cook's transformation not only shows that satisfiability is NP-complete, but also demonstrates the use of both predicate and propositional calculus as encoding languages. The great difficulty in solving the satisfiability problem in polynomial time is surely due to this encoding power. Perhaps satisfiability encodes other processes that could not possibly themselves be solved in polynomial time.

Problems

1. Convert each of the predicate clauses displayed in the generic transformation to disjunctive form.

2. Rewrite the three generic clauses involving the effects of a nondeterministic program on Turing machine operation. For q and s substitute 3 and 0, respectively. Use the program appearing in the illustration to decide what q', s', and c' must be.

3. Write an expression for the total number of literals appearing in an instance of satisfiability generated by Cook's generic transform. Explore the size of this expression when it is restricted to the definition of just one Turing machine transition.

4. Using the definition of a random access machine introduced earlier, prove a different version of Cook's theorem in which nondeterministic RAMs are the vehicle for membership in NP.

References

M. R. Garey and D. S. Johnson. *Computers and Intractability: A Guide to the Theory of NP-Completeness.* Freeman, San Francisco, 1979.

A. V. Aho, J. E. Hopcroft, and J. D. Ullman. *The Design and Analysis of Computer Algorithms.* Addison-Wesley, Reading, Mass., 1974.

46

![hand symbol]

SELF-REPLICATING COMPUTERS

Codd's Machine

Just before his death in 1957, the great John von Neumann conceived of a vast machine that was able not only to compute any computable function but also to reproduce itself! Not surprisingly, this machine's method of creating offspring had little apparent resemblance to human methods. Moreover, it existed only as an incomplete specification for how to implant an enormous collection of 29-state automata in the plane (see Chapter 2). Even if it *had* been built, it would probably not have been impressive to watch since only the pattern of its states — and not the underlying physical devices — would have been reproduced.

To be more specific, here is the sort of thing von Neumann had in mind: The plane is divided into an infinite square grid, and each square is occupied by the same automaton, say *J*. Initially, all but a finite number of these automata are in a special, quiescent state, with the remainder displaying a pattern of states repre-

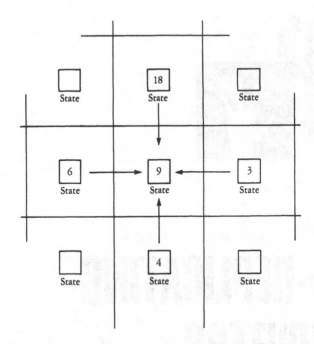

Figure 46.1 A neighborhood in the von Neumann cellular space

senting the von Neumann machine. The state of each automaton at time $t+1$ depends strictly on its own state and on the state of its four edge neighbors at time t (Figure 46.1).

Not only could the von Neumann machine compute any specified Turing-computable function (see Chapter 31), but also it could cause a duplicate pattern of states or, indeed, a pattern of states representing any specified Turing machine, to appear elsewhere on the grid. The machine thus created could then be turned loose on any computation for which it was designed.

It is amusing to visualize von Neumann's universal computer-constructor (UCC) thus propagating itself endlessly on some infinite conceptual grid!

In the mid-1960s, E. F. Codd improved the von Neumann design in many respects, the chief of which is a reduction from 29 to just 8 states. It would require most of this book just to describe Codd's machine in detail, so only a general description of the machine can be given here. However, to provide some insight into the nature of such details, a specific aspect of the eight-state machine, namely its "constructor paths," is explained.

A general picture of Codd's machine (henceforth called the UCC, short for Universal Computer Constructor) would involve a black box and a number of tapes embedded in the middle of a quiescent, one-way, infinite slot in the grid (Figure 46.2).

The box marked *UCC control and executive* consists of an immense pattern of states, continually changing when the machine is in operation; these embody something like the logic circuitry of a modern computer. This vast, planar computer operates as follows:

The description of an arbitrary Turing machine (program) is placed on the program tape, and the data on which that Turing machine is to operate, are placed on the data tape. The control and executive section then reads from the program tape and mimics the action of the Turing machine there specified by reading from and writing upon the data tape. In doing this, a data path must be created by the UCC from the control and executive block to each of the two tapes in question. The paths are extended or retracted as the simulation of the Turing machine proceeds. Patterns of states representing symbols read or to be written traverse these paths.

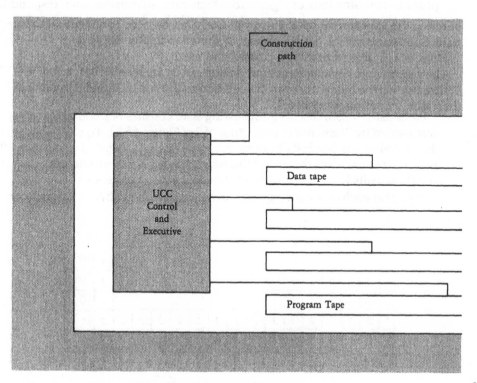

Figure 46.2 Schematic of the UCC

The UCC is also a "universal constructor." Given the description of a particular cellular Turing machine on the program tape, the UCC is able to construct that machine by extending a construction path into a quiescent area adjacent to the UCC and there setting up a pattern of states that, when activated, computes the function proper to that Turing machine. Indeed, the UCC can construct a copy of itself in precisely this manner.

The machine to be constructed by the UCC is described in a special language which, in effect, tells it how to carry out the construction in terms of the deployment of the constructor path. Specifically, to ensure that a certain target cell is placed in the correct initial state, the path is extended to that cell, and a special signal is propagated along the path to effect the appropriate transition.

Speaking of transitions, although each cell of this space contains an eight-state automaton, the alphabet is rather large. In fact, with each of four neighboring cells (and itself) being in any one of eight possible states, there are

$$8^5 = 32,768$$

possible combinations of signals to which each automaton must respond by entering the appropriate state. However, Codd has no need to specify all these transitions since only a "small" subset of them is actually needed.

Here, at last, is how paths are created and used.

The various states in which each automaton can be are labeled 0, 1, 2, 3, 4, 5, 6, 7, with 0 being the quiescent state. If a cell is in state 0 and all its neighbors are in that state, it continues in state 0.

Essentially, a path in a nonpropagating state consists of a configuration of 1s surrounded by "insulation" consisting of 2s (Figure 46.3). To extend a path in the direction it is currently headed, pattern 0607 is propagated along the path. By the time it reaches the end of the path, one cell has been added to the path and three cells have been added to the insulation (Figure 46.4).

Note that each of these snapshots involves transitions by a small number of

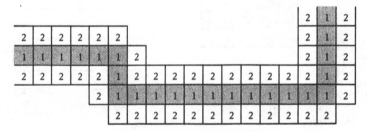

Figure 46.3 A nonpropagating path

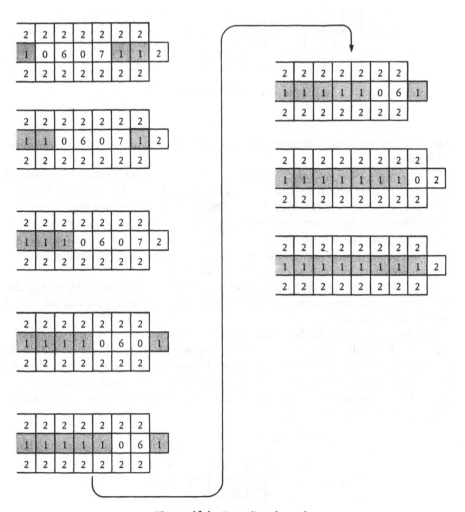

Figure 46.4 Extending the path

automata. For example, the cell with the dark outline in the second snapshot of the path is in state 1 and surrounded by cells in states 7, 2, 2, 2, taken in a conventional order. Under the transition rules governing the Codd automation, we then have the transition taking place in that cell by the next instant of time. This transition holds true of any automaton in the infinite array comprising the cellular space: When in state 1 and when its neighbors are in states 7, 2, 2, and 2, respectively, a cell must then enter state 7. The reader can examine the above

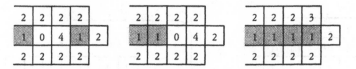

Figure 46.5 The path is about to turn . . .

path developing and write down several other transitions which the automata must all obey.

In growing a path, whether to read a tape or to extend it into the quiescent region, it is also useful to have the path turn to the right or to the left. For example, a left turn is signaled by the state pattern 04 propagating along the path (Figure 46.5). When this configuration is achieved, another 04 is sent along the path (Figure 46.6). Next, an 05 is transmitted, followed by an 06, resulting finally in the configuration shown in Figure 46.7, which can now be extended farther upward by the 0607 command.

There exist path signals which result in specific states being assigned to the cell at the tip of a construction path and other signals which result in the path's step-by-step retraction.

Although only parts of this machine have been simulated on a computer in order to check the correctness of its components, it has been specified in enough detail to actually build one, should someone with a few hundred years available be willing to set up a physical cellular array of large enough size. In any case, it is probably, so far, the largest and most complicated computing device ever conceived.

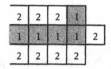

Figure 46.6 . . . left.

Figure 46.7 The turn completed

Figure 46.8 A cellular tape

If we do not require that a self-replicating computer be a universal constructor, much simpler automata become possible. For example, in 1985 an American computer scientist, Christopher Langton, discovered a self-replicating automaton that required just eight states per cell. A table listing 219 transitions governed the Langton automaton as it grew from an initial looplike configuration into a "colony." In 1989 a Canadian mathematician, John Byl, found an even simpler self-replicating machine. Byl's automaton requires a mere six states per cell, a table of only 56 entries, and an intial "seed" of 11 cells.

		2	2	
	2	3	1	
	2	3	4	2
		2	5	

The following table shows Byl's automaton as a set of 56 transitions. The first entry, CNESW → C, indicates the form followed by the transitions. The numerical states of the central, northern, eastern, southern, and western neighbors of C produce the new state for C shown to the right of the arrow.

CNESW → C	10000 → 0	20000 → 0	30003 → 0	31235 → 5	40252 → 0
00003 → 1	10001 → 0	20015 → 5	30011 → 0	31432 → 1	40325 → 5
00012 → 2	10003 → 3	20022 → 0	30012 → 1	31452 → 5	4xxxx → 3
00013 → 1	10004 → 0	20202 → 0	30121 → 1	3xxxx → 3	
00015 → 2	10033 → 0	20215 → 5	30123 → 1		50022 → 5
00025 → 5	10043 → 1	20235 → 3	31122 → 1	40003 → 5	50032 → 5
00031 → 5	10321 → 3	20252 → 5	31123 → 1	40043 → 4	50212 → 4
00032 → 3	11253 → 1	2xxxx → 2	31215 → 1	40212 → 4	50222 → 0
00042 → 2	12453 → 3		31223 → 1	40232 → 4	50322 → 0
0xxxx → 0	1xxxx → 4	30001 → 0	31233 → 1	40242 → 4	5xxxx → 2

Entries with four x's in a row mean simply "all other combinations" with the particular digit that leads the four x's.

Problems

1. Examine the different stages in the construction of a path. How many state transitions can you identify?

2. Recently, the game of Life (see Chapter 44) was shown to be computation-universal in the same sense as the UCC. Explain how gliders could be used to transmit information.

3. Write a computer program that simulates Byl's self-replicating cellular automaton. What happens to the seed units inside the ever-growing colony?

References

E. F. Codd. *Cellular Automata.* Academic Press, New York, 1968.

J. von Neumann. Theory of automata: Construction, reproduction, homogeneity. *The Theory of Self-Reproducing Automata* (A. W. Burks, ed.), pt. II. University of Illinois Press, Urbana, 1966.

STORING IMAGES

A Cat in a Quad Tree

The arrival of quad trees in the early 1970s signaled a new stage of progress in computer graphics. Many operations, whether for the storage, manipulation, or analysis of computer images, were enhanced. Entirely new operations were made possible — all by the simple expedient of subdividing a digital image by quadrants and identifying the quadrants as nodes in a tree, the quad tree.

Consider Figure 47.1. A digital cat sits in profile. The cat may be decomposed (so to speak) into a quad tree. Armed with this structure, we may store the cat image very efficiently, carry out various geometric transformations on it, and determine whether there are any objects besides the cat.

The quad tree for our cat is generated by dividing the image matrix into quadrants, then subdividing each of these into quadrants, and so on. The last stage of subdivision is reached when each pixel, i.e., picture element, becomes a quadrant in its own right. For such subdivision to work, the size of the matrix must obviously be a power of 2. But this is usually easy to arrange.

Each quadrant in the subdivision is represented by a node in the corresponding quad tree; if the quadrant is all black or all white, the node is tagged

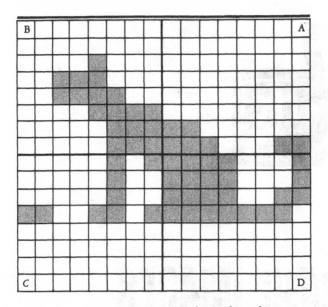

Figure 47.1 A cat and something else

accordingly and treated as a terminal node. It has no children. But if the quadrant is neither all black nor all white, the corresponding node is given four child nodes, one for each subquadrant. These are taken in the same order as the original four quadrants labeled *A*, *B*, *C*, and *D* in the illustration.

When it is decomposed by this process, the cat image is represented by a single node representing the entire matrix. The four principal quadrants are represented by four nodes appropriately labeled. The first node is the northeast quadrant, the second node is northwest, the third node is southwest, and the last node is southeast (Figure 47.2). Each of the four nodes gives rise to a subtree that is the quad tree for that quadrant; node *B*, for example, yields the subtree shown in Figure 47.3.

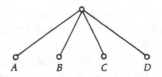

Figure 47.2 The principal nodes of a quadtree

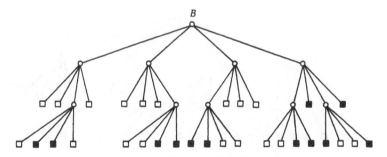

Figure 47.3   The quadtree in quadrant *B*

The quad tree above *is* the image of the cat when it is interpreted properly. A white terminal node that is k levels up from the bottom of the tree represents a $2^k \times 2^k$ quadrant that is entirely white. And so with black. The total number of nodes in the quad tree tracks the storage space it requires very closely. Thus the cat has been compressed into 41 nodes. At the same time, the cat's image in quadrant B occupies a 16×16 matrix. By conventional storage methods, 64 pixels must be stowed away. For the vast majority of images that arise in various applications, this kind of storage economy is typical.

A number of manipulations of quad trees correspond to standard manipulations of graphic images. For example, to rotate the image of the cat 90° counterclockwise, the nodes at each level must be rotated, in effect. Thus the root of the tree would display a different order (Figure 47.4). Quadrant *D* now occupies the northeast corner of the picture, *A* the northwest, and so on. The subtree pendant at *B* now takes on a different appearance as each of its nodes are rotated (Figure 47.5).

Among the other manipulations that can also be carried out on images represented by quad trees are changes of scales (by a factor of 2) and translations. To change the scale of an image, it is only necessary to remove all the terminal nodes at the bottom level of the tree and to reinterpret the root node as representing one principal quadrant. This shrinks the image by a factor of 2. To blow

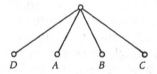

Figure 47.4   Rotating the principal nodes

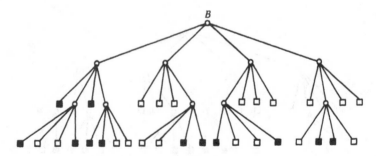

Figure 47.5   Rotating the quadtree at *B*

it up, the image must initially occupy just one principal quadrant; remove the root node and the three nodes corresponding to the quadrants not occupied by the image.

The shrinking operation is closely related to a useful approximation technique. If one or more levels are removed from the bottom of the quad tree, the resulting image loses a factor of 2 or more in resolution. It becomes approximate. For some applications such as visual reference and pattern recognition, it is sufficient to work with low-resolution images, especially ones that are so easily produced.

The final operation reviewed here identifies the components of an image. In the picture presented at the beginning of this chapter, there are two components, the cat and two adjacent pixels in front of it. It is difficult to display a convincing mouse at this resolution, but that is what the pair of pixels represents. The image, in other words, has two components. Finding the components of an image is important in numerous applications, such as the automatic scanning and counting of separate features in medical images.

The component-finding algorithm locates a black pixel and searches for black neighbors to the north, east, south, and west. It searches continually outward, assembling a list of all black pixels in the same component while continually adjoining new neighbors to the list. To identify a component in this manner, a neighbor-finding algorithm is therefore crucial. Such an algorithm operates on the quad tree, of course, so it is reasonable to speak of nodes in the tree and certain quadrants of the image matrix interchangeably. The neighbor of a given quadrant in a given direction is the smallest quadrant of at least that size sharing its border on the given side. This description is readily converted to an algorithm for traversing the quad tree. Suppose, for example, that we are searching for the western neighbor of a given node. In the algorithm that follows, the *A, B, C, D* quadrant-labeling convention introduced earlier is used. Thus, a *B* quadrant is a western neighbor of an *A* quadrant, and a *C* quadrant is a western neighbor of a *D* quadrant. As the algorithm ascends the tree, it pushes the labels it encounters onto a stack. As soon as it reaches a node via a link marked *B* or *C*, the algorithm descends, always taking a link that is horizontally opposite to the one ascended on the same level of the tree.

1. **repeat**
   1. ascend one link
   2. push its label onto stack
2. **until** link label is *B* or *C*
3. **repeat**
   1. pop label from stack
   2. descend link with opposite label
4. **until** node is terminal or stack is empty

This algorithm can be generalized to take any direction as input and to find the neighbor of a given quadrant in that direction.

To use the quad tree to identify components, the tree must be traversed in some order (see Chapter 11). As each new black terminal node is visited, it is given a new label (if it is still unlabeled), and its four neighbors are examined by using the neighbor-finding algorithm above. A black neighbor (which must be a terminal node by the definition of the neighbor) receives the same label as the node being traversed:

1. *label* ← 1
2. **for each** terminal node in traversal
   1. **if** node black and unlabeled
      1. **then** label node
      1.2. **for each** neighbor
         if neighbor black and unlabeled
            then label neighbor
            else add pair of labels to list
      1.3. *label* ← *label* + 1

This is not the whole component-finding algorithm, but it is the major part. The list of equivalent pairs it produces is actually a graph which the next part of the component-finding algorithm must investigate. The graph can be explored in depth-first fashion. As long as new pairs can be found for which one label is in an equivalence class and the other is not, the algorithm continues. Finally, there will be no pairs left to process, and each equivalence class will now receive a new label. At this point, the algorithm could output the number of such new labels (two in the cat-and-mouse example).

The third stage of the component-finding algorithm simply traverses the quad tree one more time, assigning the new label that is equivalent to each of the old labels encountered during the traversal.

Problems

1. Write an algorithm that converts a binary image matrix to the corresponding quad tree.

2. Although most images requires less storage in quad tree form than in matrix form, there is one definite exception. Find the worst case in this respect.

3. Construct a new quad tree for the cat after it has pounced on the mouse, i.e., moved two pixels west. Design a general translation algorithm for horizontal and vertical motion.

References

Hanan Samet. *The Design and Analysis of Spatial Data Structures.* Addison-Wesley, Reading, Mass., 1989.

Hanan Samet, *Applications of Spatial Data Structures: Computer Graphics, Image Processing, and GIS.* Addison-Wesley, Reading, Mass., 1989.

48

THE SCRAM

A Simplified Computer

The principles governing the design and operation of a real computer can be illustrated by the complete description of a "toy" computer. No one would seriously suggest that the computer shown in Figure 48.1 be manufactured (for one thing, its very small memory renders it useless for most practical computations). But it does have all the major features of larger and more sophisticated machines; only input and output hardware has been omitted. In short, it is a Simple but Complete Random Access Machine.

Before we plunge into the details of the SCRAM machine, a brief preview of its major components is in order. The acronym MUX stands for multiplexer (see Chapter 28). This is a device that routes information from different sources to a single destination. The choice of source depends on a control signal generated by the control logic. Seven registers appear in the diagram. They are the program counter (PC), instruction register (IR), memory address register (MAR), memory buffer register (MBR), the accumulator (AC) and the adder (AD). The last register is incorporated into the arithmetic logic unit (ALU). A timer T generates pulses that are decoded into separate input lines for various

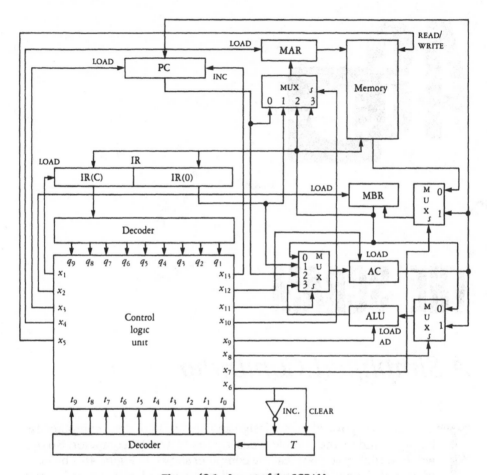

Figure 48.1 Layout of the SCRAM

destinations within the control logic unit (CLU). Another decoder translates program instructions in the IR for use by the CLU.

The diagram is entirely schematic; there is almost no relationship between its geometry and a layout of this circuit as silicon chips on a board. Indeed, the memory unit would be much larger in an actual circuit. As a general rule, long, thin rectangles represent registers, and anything else represents control or memory logic.

The SCRAM has 8-bit words, and the memory contains 16 words. It follows that 4 bits is sufficient to specify the address of any one of the memory words. As a general rule, each word of memory will contain either a number or a program instruction. The number can be up to 8 bits in length, but the instruction must share these bits between an operation code (4 bits) and an operand (4 bits). In normal 8-bit machines, operations and operands are stored in alternate memory locations, but the circuitry must be more complex. The SCRAM is too simple for such sophistication, and the result is that only 4-bit numbers can be stored in the operand for any instruction.

The heart of the SCRAM computer is the CLU. We will follow one complete cycle of operation for each instruction of the low-level programming language described in Chapter 17. These instructions are listed below.

The cycle has two parts. They are called *fetch* and *execute*. The fetch cycle gets the next executable instruction in the program currently running and loads it into the instruction register. This cycle itself is written as a special sequence of elementary machine operations called a *microprogram*. Each line of the micro-program is called a micro-operation and is written in a special replacement notation called *register-transfer language*:

t_0: MAR $\leftarrow$ PC
t_1: MBR $\leftarrow$ M, PC $\leftarrow$ PC $+$ 1
t_2: IR $\leftarrow$ MBR

All SCRAM micro-operations are set in motion by a timer T that feeds time signals via a decoder into the CLU. When the line labeled t_0 contains a 1, the contents of the PC are transferred to the MAR. In other words, the program counter contains the address in memory of the next executable instruction, and this address is transferred to the memory address register. When t_0 drops back to 0, t_1 contains a 1 and this signal initiates the next micro-operation; the contents of memory at the address contained in the MAR are transferred to the memory buffer register (see Chapter 38). At the same time, the program counter is incremented. This ensures that unless the instruction about to be executed is a JUMP command, the next instruction to be executed is the one following the current instruction in the program stored in memory. The current instruction consists of an operation code and an operand. Both are transferred at time t_2 into the instruction register from the MBR. The 4 high-order bits of the IR comprise a kind of subregister that we may call IR(C). The 4 low-order bits are called IR(0). IR(C) contains the instruction code, and IR(0) contains the operand.

The instructions, their codes, operands, and meanings are listed in the table at the top of the next page.

Operation	Code	Operand	Meaning
LDA	0001	X	Load contents of memory address X into the AC.
LDI	0010	X	Indirectly load contents of address X into the AC.
STA	0011	X	Store contents of AC at memory address X.
STI	0100	X	Indirectly store contents of AC at address X.
ADD	0101	X	Add contents of address X to the AC.
SUB	0110	X	Subtract contents of address X from the AC.
JMP	0111	X	Jump to the instruction labeled X.
JMZ	1000	X	Jump to instruction X if the AC contains 0.

The code subregister IR(C) is connected by four parallel lines (shown in the diagram as a single line) to a decoder which produces a 1 on exactly one of the nine input lines to the CLU. Each of the nine possible instruction types has its own characteristic binary pattern, and the decoder activates the appropriate line to the CLU in consequence (see Chapter 28).

The execution phase of the basic cycle immediately follows the fetch phase. At this point the t_3 input line carries a 1. At this and at subsequent times in the execution cycle, various micro-operations are performed, depending on the instruction type being executed. Each instruction has its own microprogram. For example, LDA has the following microprogram:

$$
\begin{array}{lll}
\text{LDA} & q_1 t_3\colon & \text{MAR} \leftarrow \text{IR(0)} \\
& q_1 t_4\colon & \text{MBR} \leftarrow \text{M} \\
& q_1 t_5\colon & \text{AC} \leftarrow \text{MBR}
\end{array}
$$

During the fetch cycle, all that was needed to trigger each of the three micro-operations was 1s appearing sequentially on the timing lines t_0, t_1, t_2. Now more complicated triggers are required: If IR(C) contains the LDA code 0001, the decoder converts this to a 1 on line q_1, with the other q-lines all containing 0s. The expression $q_1 t_3$ is a logical expression meaning, in effect, "if q_1 and t_3, then . . . ." We will see later how the trigger is implemented by the output of an AND gate.

Thus when q_1 and t_3 are both 1, the SCRAM loads the operand portion of the instruction register into the memory address register. At the next time step, the contents of that address in memory are loaded into the memory buffer register and then into the accumulator.

The operational cycle of the SCRAM may require up to 10 consecutive time periods. The periods are each, let us say, a microsecond in length and are generated by a timing register T. The CLU increments T (based on its own clock) between micro-operations and clears T when a new operational cycle is to begin.

The next example of microcoding of program instructions is more complicated. The LDI command requires five micro-operations:

LDI X: $q_2 t_3$:   MAR $\leftarrow$ IR(0)
            $q_2 t_4$:   MBR $\leftarrow$ M
            $q_2 t_5$:   MAR $\leftarrow$ MBR
            $q_2 t_6$:   MBR $\leftarrow$ M
            $q_2 t_7$:   AC $\leftarrow$ MBR

The only difference between direct and indirect LOAD commands is that the latter requires an additional set of memory micro-operations. This is because indirection requires two separate memory retrievals, the first to get the address for the second. Shown below are the micro-programs for two other program instructions. Readers should not find it difficult, after this, to construct satisfactory microprograms for the remaining instruction types.

ADD X: $q_5 t_3$:   MAR $\leftarrow$ IR(0)
            $q_5 t_4$:   MBR $\leftarrow$ M
            $q_5 t_5$:   AD $\leftarrow$ MBR
            $q_5 t_6$:   AD $\leftarrow$ AD $+$ AC
            $q_5 t_7$:   AC $\leftarrow$ AD

JMP X: $q_7 t_3$:   AC $\leftarrow$ PC
            $q_7 t_4$:   AD $\leftarrow$ AC
            $q_7 t_5$:   AC $\leftarrow$ IR(0)
            $q_7 t_6$:   AD $\leftarrow$ AD $+$ AC
            $q_7 t_7$:   AC $\leftarrow$ AD
            $q_7 t_8$:   PC $\leftarrow$ AC

Note that the ADD command first retrieves the number stored at memory address X and then loads it in the special arithmetic register AD (not shown in the diagram). After the contents of the accumulator are added to the AD, the result is placed back in the AC.

The second microprogram uses the same operation to increment the program counter. This register contains the address of the next program instruction to be executed. If the current instruction is not a JUMP, the CLU will merely increment the PC sometime during its operational cycle.

At this point the question naturally arises of just how the nine different microprograms are implemented in actual logic circuits. Here again, we follow time-honored pedagogic practice by employing standard logic gates for the purpose. In any event, it is a relatively simple exercise to convert these circuits to any logically complete set of gates (see Chapter 3).

It is a relatively easy matter to implement the fetch cycle and the nine possible versions of the execution cycle in individual logic circuits.

The fetch cycle is shown in Figure 48.2. For the time being, we pretend that the line x_{10} is doubled; after all, x_{10} operates a multiplexer (MUX) with two control inputs. When t_0 is 1, both x_{10} lines are to be 0, causing the MUX to select input line 0 from the PC for transmission to the MAR. The x_4 line causes the MAR to be loaded in accordance with a register design appearing in Chapter 38. When t_1 is 1, x_7 is 0 and x_5 and x_{13} are both 1. This means that the MUX serving the MBR selects memory input, the read/write line to memory carries a read (0) signal, and the PC is incremented. When t_2 is 1, the instruction register is loaded with the current contents of the MBR.

A circuit that implements the LDA instruction is shown in Figure 48.3.

When it is considered in isolation, the operation of this circuit is also straightforward. First, the paired x_{10} lines control the MUX selecting input for the MAR. When t_3 and q_1 are 1, x_{10} is 01. When t_4 is 1, the x_7 line sends a 0 control signal to the MUX selecting input to the MBR. In this case, 0 means memory. Finally,

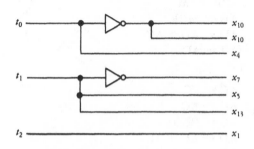

Figure 48.2 Logic for the fetch cycle

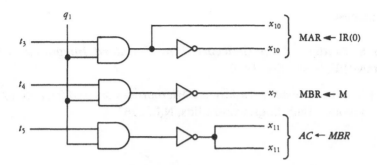

Figure 48.3 Logic for loading the accumulator

when t_5 is 1, both control lines to the MUX selecting input for the AC are 0. In this case 00 means that the MBR is selected.

Readers may have noticed a problem at this point; some of the output lines just discussed appear in both circuits. Since outputs of two different gates cannot be tied directly together, more gates must be used to effect this purpose. (See the problems that follow.)

Problems

1. Write microprograms for STA, STI, and JMZ. Implement the microprograms in standard logic.

2. Design that portion of the CLU that determines the two output lines labeled x_{10}. Input to this subcircuit will be one or both of the lines previously labeled x_{10} in the individual circuits for LDA, LDI, and the circuits designed in Problem 1.

3. Convert the following program to the equivalent set of binary words, as indicated in this chapter. This is called *machine code*. Trace the execution of the program by listing the q, t, and x variables across the top of a sheet of paper. For each step in the SCRAM's operation, fill in one line of the table by writing down the value of each input and output variable:

    LDA 1
    ADD 2
    STA 3

References

Richard S. Sandige. *Data Concepts Using Standard Integrated Circuits.* McGraw-Hill, New York, 1978.

Ronald J. Tocci and Lester P. Laskowski. *Microprocessors and Microcomputers,* 2d ed. Prentice-Hall, Englewood Cliffs, N.J., 1982.

SHANNON'S THEORY

The Elusive Codes

ny message consisting of words can be encoded as a sequence of 0s and 1s. The latter symbols can be transmitted by wire, radio waves, or a variety of other means. In all cases the message is liable to corruption. A 0 can inadvertently be changed to a 1, or vice versa. Whatever the actual source of interference or "noise," as information theorists call it, the tendency of a bit to change can be laid at the doorstep of a creature I call the noise demon With a certain probability, say p, the noise demon alters each bit being transmitted (Figure 49.1).

One way to fool the demon is to transmit three 0s for each 0 intended. Also, replace a 1 by three 1s. Assuming the receiver is in synchrony with the sender, so that the start of each triad is known, the decoding rule is very simple, as shown in the table at the top of the next page. What is the probability of a message being corrupted under this scheme? It is the probability that at least 2 bits out of 3 get changed. Thus if 000 is sent, it can become 110, 101, 011, or 111 with respective probabilities p^2q, pqp, qp^2, or p^3, where $q = 1 - p$. Adding these probabilities yields the formula $3p^2 - 2p^3$, which means that if p is less than .5, the new scheme of triples guarantees a much lower probability of the demon's

Received	Means
000	0
100, 010, or 001	0
011, 101, or 110	1
111	1

success. For example, if $p = .1$, then the probability that the demon will corrupt the message irretrievably is .028.

The foregoing coding scheme used simple redundancy as a guard against errors. Other, more sophisticated methods are available (see Chapter 12). For example, one may group the message bits into pairs and transmit the pairs along with 2 extra check bits according to the following scheme.

Message Bits	Check Bits
$a_1 a_2$	$a_3 a_4$
0 0	0 0
0 1	1 1
1 0	0 1
1 1	1 0

The first check bit, a_3, simply echoes the second message bit a_2; $a_3 = a_2$. The second check bit is called a *check sum*. It is the logical sum of the first 2 bits; $a_4 = a_1 \oplus a_2$.

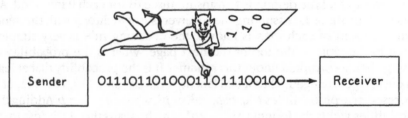

| Sender | 0111011010001101110100 ⟶ | Receiver |

Figure 49.1 The noise demon at work

There is no guarantee for any coding scheme that the corruption demon will fail. If it succeeds in changing enough bits, no decoding scheme will be able to recover the original message. For this reason, the message is decoded by the maximum-likelihood method. For each received string of 4 bits, what is the most likely interpretation of the first 2 bits? The following algorithm takes the 4 received bits b_1, b_2, b_3, b_4 as input and outputs b_1 and b_2, altered according to probable errors detected by the algorithm:

1. **input** b_1, b_2, b_3, b_4
2. **if** $b_4 \neq b_1 \oplus b_2$
     then if $b_3 \neq b_2$
          then $b_2 \leftarrow \bar{b}_2$
          else $b_1 \leftarrow \bar{b}_1$
3. **output** b_1, b_2

The operation of the decoding algorithm is illustrated in the hypercube diagram in Figure 49.2. Each vertex represents a possible 4-bit string to be received. Four of the vertices are circled. These represent the four original (probably uncorrupted) code words.

Vertices representing strings that differ in only 1 bit are connected by an edge. If one analyzes the action of the algorithm on each of the 16 possible received words, one discovers that each word is reinterpreted as the "nearest" code word. The concept of distance applied here is the Hamming distance, the minimum number of edges traversed in going from one vertex (possible word)

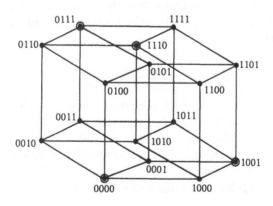

Figure 49.2 An error-correcting code in a hypercube

to another. The Hamming distance is clearly the number of bits altered in the process. Maximum-likelihood decoding amounts to selecting, for each possible message word, a code word that is a minimum Hamming distance from the message word.

The encoding/decoding scheme just described enables the user to escape from all but four possible single errors: four vertices in the diagram above each have Hamming distance 1 from two code words. It is therefore possible for the algorithm to be fooled when the noise demon creates a single error in the message. But the probability of this happening is just $\frac{1}{6}p$ since only 2 of the 12 messages containing a single error will be misinterpreted, on average. In a few cases when two or more errors occur, the algorithm can still recover the original message. It follows that the probability that the algorithm is fooled by the noise demons is actually less than $\frac{1}{6}$. When $p = .1$, this probability becomes less than .016, which is substantially better than the previous scheme.

In the first example, we sent 1 message bit and 2 check bits. The probability of falling afoul of the noise demon was $3p^2 - 2p^3$. In the second example, we sent 2 message bits and 2 check bits. The probability of irrecoverable errors was less than $p/6$ plus the probability of two or more errors, namely, $3p^2 - 2p^3$. But in the second case 2 message bits were transmitted, and therefore the recovery probability per bit was less than $p/12 + 3p^2/2 - p^3$. In the case where $p = .1$, this becomes .0198, a definite improvement.

There are clear advantages to sending code words that are relatively long. A fundamental theorem discovered by Claude Shannon of Bell Laboratories in 1948 makes the advantages explicit. Unfortunately, the theorem gives no hint of how to construct codes that exploit the advantage!

In what follows, we assume that the codes under discussion have words of length n. The channel demon will busily subvert messages with a uniform probability of p per bit. Suppose that code C has m words $X_1, X_2, \ldots, X_m$ and that a maximum-likelihood decoding algorithm for it incorrectly decodes the word X_i with probability p_i. Then the probability of incorrectly decoding the average word in C is

$$\frac{1}{m} \sum_{i=1}^{m} p_i$$

Denote by $P(n)$ the minimum such probability over all codes C of size m and length n. Shannon's fundamental theorem describes conditions under which this probability approaches 0 as n approaches infinity. Specifically, the *information rate r* of a code is simply $\log_2 m$ (the amount of information embodied by the set of all code words) divided by the length n of the code. If r is greater than 0 and less than $p \log p + q \log q$ and if $m = 2^{nr}$, then

$$P(n) \rightarrow 0 \text{ as } n \rightarrow \infty$$

In words, Shannon's theorem tells us that for any small positive number ϵ, there is a word length n and a code C having that word length such that the probability of a decoding failure is less than ϵ! The theorem is generous in its allowance of words, moreover; the number available is exponential in n.

The proof of Shannon's theorem is somewhat lengthy and technical, but the main idea is easily described. The proof proceeds by selecting m code words at random from the set of all binary words of length n. The latter set constitutes a kind of space in which the distance between two words is the Hamming distance. In the proof of Shannon's theorem, each word in the randomly selected code is surrounded by a "sphere" of radius p. This is just the set of all n-bit words having a Hamming distance of p or less from the code word acting as the sphere's center.

The decoding rule used in the proof is a strict form of maximum-likelihood decoding. When a word is received, find the code word that is closest in terms of Hamming distance. If the received word lies within p of the computed code word, select the latter as the intended message. Otherwise, declare a decoding failure. If p is chosen to be a certain simple polynomial in n, it turns out that the probability of a received word lying outside its proper p sphere is arbitrarily small, at least for large enough n. Thus the probability of a decoding failure by the maximum-likelihood method is at least as small: it tends to 0 as n goes to infinity.

But why, then, is it so difficult to find good codes? Why not choose one at random? The answer lies in the size that n must be before the desired failure probability ϵ is reached. It is simply too large to be practical. At the same time, the proof of Shannon's theorem only addresses the average code in the space of all n-bit words. It merely guarantees that one can do at least as well as ϵ with n words. Perhaps a much smaller n will suffice. It frequently does, but shorter codes must be found by other methods that will continue to challenge the ingenuity of theoreticians for decades to come.

Problems

1. Suppose a 3-bit message $a_1 a_2 a_3$ is to be transmitted via a noisy channel. Adding 3 check bits $a_4 a_5 a_6$ that are computed by the following scheme results in a 6-bit code word to be transmitted.

$$a_4 = a_2 \oplus a_3$$
$$a_5 = a_1 \oplus a_3$$
$$a_6 = a_1 \oplus a_2$$

Devise an algorithm that recovers the most likely uncorrupted message.

2. Write a program that takes two integers m and n as well as a real number p as input. It generates m random n-bit code words and enters them in an array. The program then systematically selects a random word from the array and alters its bits with probability p, thus imitating the action of a noise demon. Doing this 1000 times, it computes the nearest code word and compares the result with the originally selected code word. The number of mismatches is converted to an estimate of decoding failure probability. Investigate how large n must be for each of a number of p values that steadily decrease in size. How fast does n seem to be growing?

References

R. W. Hamming. *Coding and Information Theory.* Prentice-Hall, Englewood Cliffs, N.J., 1980.

W. W. Peterson and E. J. Weldon, Jr. *Error-Correcting Codes.* M.I.T. Press, Cambridge, Mass., 1981.

50

DETECTING PRIMES

An Algorithm that
Almost Always Works

No one has yet devised an algorithm for deciding in polynomial time whether a number is prime. Is there a polynomial p and an algorithm A such that, for any positive integer n, A can discover whether n is prime in just $p(\lceil \log n \rceil)$ steps? Here, we use $\lceil \log n \rceil$ as a measure of input length since it is assumed that n is being represented in binary notation.

As with many seemingly intractable problems, as soon as the size of the input becomes moderately large, currently known algorithms for solving them simply take too long. For example, no one has the patience to wait while such an algorithm tries to decide whether a number in the neighborhood of 2^{400} is prime. It would seem that for some problems, at least, this is the price we must pay for absolute certainty about the answer—when it does come!

Imagine, then, an algorithm which, in the space of *a few minutes*, decides that

$$2^{400} - 593$$

is prime with probability

$$.99999999999999999999999999999999999999999999999$$
$$99999999999999999999999999999999999999999999999$$
$$99999999999999999999999999999999999999999999999$$
$$99999999999999999999999999999999999999999999999$$
$$99999999999999999999999999999999999999999999999$$
$$99999999999999999999999999999999999999999999999$$
$$99999999999999999999999999999999999999999999$$
$$99999999999999999999999999999999999999999999$$

An algorithm like this, especially if it were easy to understand and to program, would have both practical and aesthetic appeal.

In fact, such an algorithm does exist. It was discovered by Michael O. Rabin and depends on the notion of integers being a "witness" to the compositeness of a number n. If a single witness can be found, n stands condemned as a composite number; but if a reasonable time is spent searching for witnesses and none can be found, then n is exalted to primality and continues to enjoy that status with little fear of ever being dethroned.

Rabin defined a *witness to the compositeness of* n to be any integer w satisfying either of the following two conditions:

1. $w^{n-1} = 1 \bmod n$, or
2. For some integer k,

$$m = \frac{n-1}{2^k} \text{ is an integer and } 1 < gcd(w^n - 1, n) < n$$

It is easy to see that the existence of a witness means that n is composite because, in condition 2, a greatest common divisor (gcd) of n (and any other number) is certainly a factor of n. It is also easy to devise an algorithm which will check in polynomial time whether a given integer is a witness to the compositeness of n.

Since witnesses are, indeed, witnesses and since it may be determined very quickly whether a given integer is a witness, it remains to ask how common they are. Certainly, if witnesses are very rare, we could hardly use our inability to find one as a basis for claiming that n is prime. This is where a theoretical result by Rabin comes in handy:

Theorem: If n is a composite number, then more than half the numbers in the set $\{2, 3, \ldots, n-1\}$ are witnesses to the compositeness of n.

It is now a simple matter to outline an algorithm to test whether n is prime:

1. Input n.
2. Select m integers $w_1, w_2, \ldots, w_m$ at random from the set $\{2, 3, \ldots, n-1\}$
3. For $i = 1, 2, \ldots, m$, test whether w_i is a witness.
4. If none of the m integers are witnesses, output yes, else, output no.

Since more than half of the integers in the set $\{2, 3, \ldots, n-1\}$ are witnesses to the compositeness of n, in case n is not prime, the probability that none will be chosen by a random selection is only $(\frac{1}{2})^m$. In other words, the algorithm has only a small chance of failing to detect the compositeness of n, especially if m is chosen to be large. Therefore, if the algorithm outputs a yes, the a priori probability that n is prime is clearly at least

$$1 - (\tfrac{1}{2})^m$$

In the example used to introduce this note, a value of $m = 400$ was used. When the number of witnesses selected for a trial has about the same order of magnitude as the size of the problem, namely $\lceil \log n \rceil$, the algorithm can be expected to output yes or no in reasonably short order. An interesting experiment run by Rabin involved a test of all numbers of the form $2^p - 1$ for $p = 1$, $2, \ldots, 500$, with only $m = 10$ witnesses in each case. The probability of error was just slightly less than $1/1000$, and in the space of a few minutes, the algorithm was done; its yeses coincided *exactly* with a table of Mersenne primes (primes of the form $2^p - 1$)!

Of course, there is always that slightly unsettling feeling after having run Rabin's algorithm on a number n and discovered that it is prime: is it really? One might be tempted to increase k to the point where the time it takes to confirm that n is prime is several hours instead of just minutes. A reasonable stopping rule has been suggested; quite moderate values of k are sufficient to guarantee that the *computing hardware* is far more likely to fail than the algorithm!

As interesting and attractive as Rabin's approach to intractable problems may seem, it is worthwhile to sound a cautionary note: One may easily invent witnesses for most kinds of problems. They may, however, be rare, and even if they are relatively common, it may be quite difficult to prove that they are.

Problems

1. A dark room contains two barrels. The first barrel is filled with orange-flavored jellybeans, the second is filled with a half-and-half mixture of orange

jellybeans and licorice-flavored ones. How many jellybeans must one eat to distinguish the barrels with a 99 percent chance of being right?

2. How does condition 1,

$$w^{n-1} \equiv 1 \bmod n$$

contribute to w's status as a witness to the primality of n?
(*Hint:* Examine condition 2.)

3. Devise a polynomial-time algorithm for deciding whether an integer w is a witness to the primality of n. The instance size must be taken as the number of digits in w.

References

M. O. Rabin. Probabilistic algorthims. *Algorithms and Complexity, New Directions and Recent Trends.* (J. F. Traub, ed.). Academic, New York, 1976, pp. 21–39.

Donald E. Knuth. *The Art of Computer Programming*, vol. 2. Seminumerical Algorithms. Addison-Wesley, Reading, Mass., 1967.

51

![Chapter 51 icon]

UNIVERSAL TURING MACHINES

Computers as Programs

I n Chapter 31, Turing machines were described as a kind of computer, albeit in abstract form. Such terminology can be misleading, however, since Turing machines are normally defined for solving specific problems and are not programmable in the usual sense of the word. Of course, if one thinks of a Turing machine only in terms of its "hardware," that is, its tape, read/write head, finite control unit, and other bits and pieces, then it is reasonable to distinguish between a Turing machine and the programs which run on it. Such programs could be inputted to the Turing machine as a list of quintuples.

In one sense, a programmable Turing machine already exists, at least in the abstract. It is called a *universal Turing machine,* and it was first defined by Alan Turing in the 1930s as an abstract model of the most general type of computing device which he could imagine.

A universal Turing machine (Figure 51.1) has a fixed program permanently

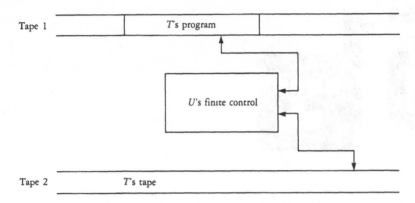

Figure 51.1 A universal Turing machine

embedded in its finite control unit. This program mimics the action of an arbitrary Turing machine (or program) by reading that program on one tape and simulating its behavior on another.

The universal Turing machine defined here will be called U and will have two tapes. Tape 1 will contain a list of the quintuples defining some Turing machine T, and tape 2 will be blank. On the latter tape, U will mimic the action of T. The initial and final appearances of tape 2 will be identical to its appearance had T been operating on it directly.

As we have already seen in Chapter 31, each two-tape Turing machine can be simulated by some one-tape Turing machine, and this observation applies to universal machines with equal force. In other words, given the construction of U in this chapter, it would be a relatively straightforward matter to construct a one-tape universal Turing machine U^1.

The fixed program inhabiting U's control section is really analogous to interpreter software in a modern digital computer. It is divided into two sections:

1. Given T's current state and input symbol, find the quintuple (q, s, q', s', d) in the description of T that applies.

2. Record the new state q', write the new symbol s' on tape 2, move the read/write head 2 in direction d, read the new symbol on tape 2, and record it beside q'.

The universal machine U expects a particular format for the description of T's program. The quintuples of T and, indeed, all T's tape symbols will use the binary alphabet. Thus, if T has n states, then k-bit binary numbers will be used to index the states of T, where $k = \lceil \log n \rceil$. Given just two tape moves, namely left

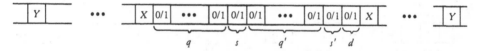

Figure 51.2 A program to be executed

(=0) and right (=1), each quintuple can be listed as a $(2k+3)$-bit binary number. The quintuples are separated by X's, and two Y's are placed as boundary markers at either end of T's quintuples. In the quintuple shown in Figure 51.2, the bits marked q and s represent T's current state and input symbol, q' and s' represent T's next state and new symbol, while d represents the direction T's read/write head is to move.

Figure 51.3 shows a portion of U's interpretive program displayed as a state-transition diagram. This diagram has been made very compact by designating each state as either right-moving (R) or left-moving (L); every transition out of such a state (and, possibly, back into it) involves a movement of U's read/write head in the corresponding direction. Many transitions undergone by U in this and subsequent diagrams are not shown. A "missing" transition simply means that U stays in the current state, moving according to that state's direction and writing whatever it reads. The purpose of the portion of U shown in Figure 51.3 is to locate the next quintuple to be executed by finding the first two items, the state q and the input symbol s.

Machine U uses the $(k+1)$-bit segment of tape 1 immediately to the left of the left-hand Y marker as a workspace in which to record T's current tape and input symbol. Given that a state label q and input symbol s already occupy this space, the q/s location program is easy to describe. With read/write head 1 on

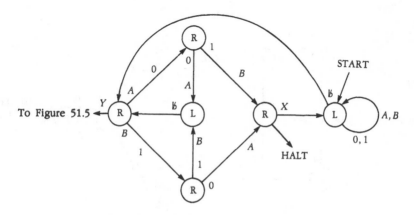

Figure 51.3 The search for the current quintuple

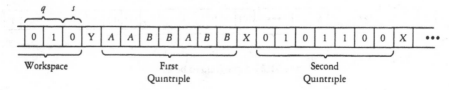

Figure 51.4 *U* locates the quintuple.

the left-hand *Y* marker, *U* begins in the state labeled START above. It scans to the left, changing each 0 to *A* and each 1 to *B* until it encounters a blank. Then it moves to the right, changing the first nonblank symbol it encounters to a 0 or 1 depending on whether that symbol was an *A* or a *B*. It then moves to the right, looking for a match for that symbol. The first 0 or 1 bit encountered triggers either a return to the same matching cycle or a return to the starting state, depending on whether a match was found. In all probability, the *q/s* portion of the first symbol will not match the *q* and *s* stored in the workspace. In such a case, *U* moves right to the first *X* and then reenters the starting state, where it changes all the 0s and 1s between that *X* and the left end of the workspace to *A*'s and *B*'s, respectively. Then *U* attempts to match the *q* and *s* in its workspace with the next quintuple. Ultimately, *Q* either succeeds, taking a transition to the next subdiagram, or halts because it never found a match. In such a case, *T* would also halt under the usual convention for Turing machines. Figure 51.4 shows tape 1 after *U* has found a match for *q* and *s* at the second quintuple.

The second portion of *U*'s program records the new state *q'* in the workspace and moves read/write head 2 according to *T*'s current quintuple. It then records the new symbol in the workspace beside *q'* on tape 1. In Figure 51.5, the square

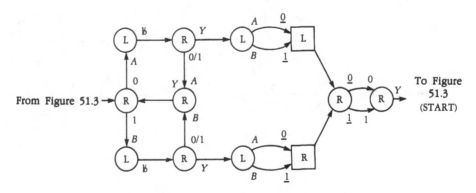

Figure 51.5 Recording the new state and moving *T*'s read/write head

state symbols indicate moves of read/write head 2. Correspondingly, under-lined 0s and 1s indicate symbols read from or written on tape 2.

This portion of U's program is invoked when the first portion has located a quintuple whose q/s component matches the workspace contents. That compo-nent has already been rewritten in terms of A's and B's, but the $q'/s'/d$ compo-nent still exists as a string of 0s and 1s. This portion of U's program begins with read/write head 2 just to the right of the left-hand Y marker.

Scanning to the right, the first binary symbols which U finds belong to the next state q'. These symbols are copied one at a time, in terms of A's and B's, into the workspace. When this section of U's program has finished copying q', it next encounters the single binary symbol for s' in T's quintuple. This is the symbol to be written by T on its tape. Machine U copies s' as an A or a B into the last (rightmost) cell of the workspace and scans rightward to the quintuple to pick up the last remaining symbol, d.

Shown in Figure 51.6 is tape 1 at this stage in U's operation. When U returns to the workspace, however, it finds it has run out of room: it encounters the Y marker instead. At this point, U "remembers" the value of d by being in the upper branch or the lower branch of its state diagram. In either case, U scans to the left in order to pick up the s' symbol as an A or a B, converts this to the corresponding 0 or 1, and then writes it on tape 2, just as T would do. Finally, U moves read/write head 2 to the left (upper branch of the diagram) or to the right (lower branch), just as T would do, and then reads the next symbol from tape 2. Replacing s' in the workspace, U shifts right to the Y marker and moves back into the portion of program depicted in the first figure.

To set U in motion at the beginning of its execution of T's program, it is necessary to place read/write head 1 over the left-hand Y. Read/write head 2 is placed over the initial cell of T's tape. The workspace must also be initialized by placing the k-bit string for state q_0 (T's initial state) along with the symbol initially scanned on tape 2. And U's program does the rest!

Turing's thesis strikes a close parallel with Church's thesis (see Chapter 66) in declaring that anything which one could reasonably mean by an "effective procedure" is captured by a specific computational scheme, in ths case, Turing machines. A universal Turing machine embodies Turing's thesis in one stroke, so to speak, by simultaneously representing all possible Turing machines and

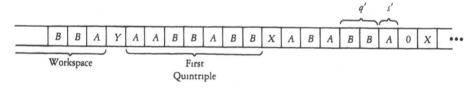

Figure 51.6 Which way will T's read/write head move?

(by Turing's thesis) all possible effective procedures. It does this on an abstract level in much the same way as a general-purpose digital computer represents all possible programs for it. Its very existence challenges us to explore the range of possible programs; what can it do and what can it not do?

In the theory of computation, the notion of a universal Turing machine serves as a focus for some questions concerning the existence of effective procedures. For example, the halting problem (see Chapter 59) asks for an effective procedure which decides, for each possible Turing machine-tape pair (T, t), whether T ultimately halts on t. Rather than having to construct such a procedure for all pairs (T, t), it is only necessary to construct one for (U, t), since t may initially contain T and therefore U will halt if and only if T does.

Problems

1. As far as T's computation is concerned, U is incredibly slow: For each state transition undergone by T, U undergoes a great many. How many in the worst case? Assume that T has n states and m quintuples in its program.

2. In Marvin Minsky's classic text (referred to below), there is a construction of a one-tape universal machine. Naturally, Minsky's machine is rather more complicated than the one appearing here since a great deal of bookkeeping is required with a one-tape restriction. How much simpler can one make U's state-transition diagram by allowing U to have three tapes?

References

Harry R. Lewis and Christos H. Papadimitriou. *Elements of the Theory of Computation.* Prentice-Hall, Englewood Cliffs, N.J., 1981.

M. Minsky. *Computation: Finite and Infinite Machines.* Prentice-Hall, Englewood Cliffs, N.J., 1967.

52

TEXT COMPRESSION

Huffman Coding

The two major applications of coding techniques are protection and compression. When a message is encoded as a string of 0s and 1s, there are techniques by which the insertion of extra bits in the string to be transmitted enables the receiver of the message to discover which 0s or 1s (if any) are in error. In addition, there are methods for shortening the string so that not as many bits need to be transmitted.

The two major spheres of application in coding theory are *time* and *space*. Time applications are the traditional ones in which a message is transmitted electronically, over time, and the concern is to protect the message from errors (see Chapter 12). Space applications involve the protection or compression of data during storage in some electronic medium such as computer memories or magnetic tapes.

If a great deal of text must be stored or if storage space is at a premium, it pays to compress the text in some manner before storing it. Huffman coding does this by exploiting redundancy in the source text.

In essence, the idea is very simple. Suppose that we have a text string using an alphabet of symbols $s_1, s_2, \ldots, s_n$ and that the probability of the ith symbol appearing at a randomly selected spot in the string is p_i. Each symbol s_i will be replaced by a binary string of length l_i so that the average string length representing a symbol will be.

$$L = \sum_{i=1}^{n} p_i l_i$$

Assuming that $p_1 > p_2 > \cdots > p_n$, it is not hard to see that L is minimized only if $l_1 < l_2 < \cdots < l_n$. This observation forms the basis for a technique that selects the actual binary strings which encode the various symbols. Suppose, for example, that the symbols A, B, C, D, E, F, and G appear with probabilities .25, .21, .18, .14, .09, .07, and .06, respectively. Note that these probabilities sum to 1.

A *Huffman tree* for these symbols and their probabilities is a convenient visual representation of the selection in which binary strings represent the symbols (Figure 52.1). Each node of the tree is labeled with the sum of probabilities assigned to the nodes below it.

Before going on to show how this tree was constructed, we note that each symbol in the source alphabet lies at a terminal node of the tree and that the

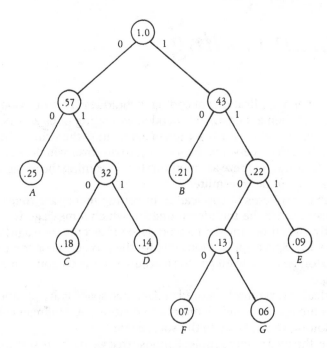

Figure 52.1 A Huffman tree

binary string which encodes it is the sequence of 0s and 1s we encounter in traversing the tree from the root to that node. The Huffman code for the system just described is therefore

A	00
B	10
C	010
D	011
E	111
F	1100
G	1101

The average code length L is 2.67 bits, and this is a minimum.

The algorithm which assigns these codes constructs an explicit Huffman tree in bottom-up fashion. An array called *subtrees* contains the index numbers of the roots of all the current subtrees from which the Huffman tree is being constructed. Initially, *subtrees* contains nothing but single-node subtrees, one for each of the n symbols s_i of the alphabet being encoded. There is, therefore, a corresponding probability array *prob* that, for each root node in each subtree under construction, contains the probability associated with that node. Two pointer arrays, *leftlink* and *rightlink*, contain the structure of the Huffman tree, and when the algorithm is finished, there will be $2n - 1$ nodes all but n of which will have pointers.

At the ith iteration of the algorithm, the first $n - i + 1$ subtrees are reordered so that their associated probabilities form a decreasing sequence. The appropriate binary digits are then added to the terminal nodes of the last two subtrees in the reordered sequence, and these are merged to form a new subtree and placed in the *subtrees* array.

```
for i ← 1 to n − 1
    reorder subtrees from 1 to n − i + 1
    add 0 to terminals of subtree n − i
    add 1 to terminals of subtree n − i + 1
    prob(n + i) ← prob(subtrees(n − i) + prob(subtrees(n − i + 1))
    rightlink(n + i) ← n − i + 1
    leftlink(n + i) ← n − i
    subtrees(n − i) ← n + i
```

The process indicated by the phrase "add 0 to terminals of subtree $n - i$" simply involves traversing the $(n - i)$th subtree and adding a 0 to each word

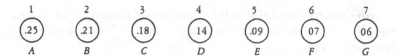

Figure 52.2 Seven single-node subtrees

stored at its terminal nodes. These words are not explicitly referred to in the algorithm but are easily handled when the algorithm is converted to a program. To "add 1 to terminals . . ." means much the same thing. After the pointers are set up associated with the new node $n + i$ (the old nodes being $1, 2, \ldots, n$, $n+1, \ldots, n+i-1$), the algorithm finally replaces the last two entries of the *subtrees* array by the root of the newly constructed subtree. When i arrives at $n-1$, there is only one subtree in the whole array, and this is the Huffman tree.

The first few steps of the algorithm's operation may be illustrated by the foregoing example. Initially, there are $n-i+1 = 7$ subtrees consisting of one node each. These are arranged in order of decreasing probability (Figure 52.2). After 0 and 1 are added to the terminals of subtree 6 and subtree 7, a new subtree with index $n + i = 8$ is created. It is linked on the right to subtree 7 and on the left to subtree 6. At the next iteration of the algorithm, the subtrees are rearranged in order of decreasing probability (Figure 52.3). At the next iteration, a new subtree is formed from two previous ones; the total number of subtrees has shrunk to 5 (Figure 52.4). After $n = 7$ iterations, there is only one subtree left, namely, the Huffman tree. At each stage code digits are assigned to the symbols lying in subtrees with the smallest probabilities. This ensures that the longest code words will tend to be used least often.

To store a given piece of text by using Huffman coding, a table is created in which the number n_i of times that the ith symbol occurs is counted. The probability p_i is then just ni/l, where l is the total length of the text. A call to the procedure above results in the assignment of a code word to each symbol, as well as in the construction of a Huffman tree. The text is scanned and, in a single

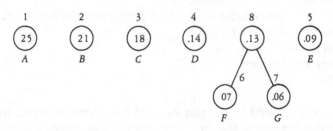

Figure 52.3 After the first iteration

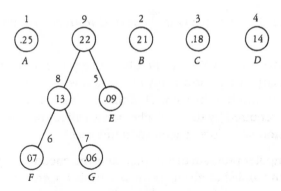

Figure 52.4 After the second iteration

pass, converted to a large binary string of concatenated code words. Next, these are divided into blocks of length m, where m is the word size of the particular computer on which the algorithm is implemented. Each block is then converted to a single integer and stored in memory in this form. The Huffman tree is also stored somewhere.

To retrieve the text thus stored, the reverse procedure is invoked, and the Huffman tree is used in the final stage of processing to extract the symbols from the binary string. Each symbol will correspond to a unique substring determined by a single root-to-terminal node search of the tree.

To avoid constructing a tree each time some text is to be stored, it may be possible to construct a more general tree, once and for all, which reflects the various symbol probabilities in a more general setting. For example, the average English sentence (ignoring punctuation) contains the symbols A to Z (and blank ♭) at a random location with the following approximate probabilities.

A	B	C	D	E	F	G	H	I
.065	.013	.022	.032	.104	.021	.015	.047	.058

J	K	L	M	N	O	P	Q	R
.001	.005	.032	.032	.058	.064	.015	.001	.049

S	T	U	V	W	X	Y	Z	♭
.056	.081	.023	.008	.018	.001	.017	.001	.172

1. A special text on a 10-symbol alphabet is to be stored in binary form by using Huffman coding. The symbols have probabilities .02, .03, .04, .06, .07, .08,

.09, .11, .20, and .30 of occurring in the text. Construct a Huffman tree for this alphabet by hand.

2. Assuming that each symbol in Problem 1 requires 4 bits to be encoded directly into binary strings, how many bits would you expect to save on a text consisting of 1000 symbols by using Huffman coding instead? The ratio of the number of bits required by the two schemes is called the *compression ratio*. What compression ratio do you achieve in this case?

3. Write a computer program that implements the encoding algorithm of this chapter using the alphabet given above as a basis for encoding English sentences. Design an experiment to compare the theoretical and actual compression ratios.

4. Write an algorithm that converts an arbitrarily large array of length-n binary strings back to the original text. There are two processes to consider: One process reads a string, watches for its end, and fetches the next word. The second process traverses a fixed Huffman tree in accordance with the stream of 0s and 1s given to it by the first process. This process must output an alphabetic symbol once it is recognized, interrupting the first process when it does.

References

R. W. Hamming. *Coding and Information Theory.* Prentice-Hall, Englewood Cliffs, N.J., 1980.

T. A. Standish. *Data Structure Techniques.* Addison-Wesley, Reading, Mass., 1980.

53

DISK OPERATING SYSTEMS

Bootstrapping the Computer

B etween a living program and the lifeless computer it must run on lies the
operating system. In the 1950s and even in the 1960s it was common to
see programmers toggling in their programs by hand, setting switches at
a computer console. Since then a multifaceted collection of programs
known as the operating system (OS) has evolved. Its presence means, among
other things, that no one has to toggle programs in by hand. The operating
system of a disk-based microcomputer, called a *disk operating system* (DOS),
illustrates the general structure and purpose of operating systems (Figure 53.1).

A computer's operating system is really a coordinated ensemble of special
programs written in machine language or in C, a special high-level language
developed especially for this purpose. The programs manage memory and
oversee the central processing unit (CPU). They run input and output devices,
provide special user commands, and generally supervise every aspect of a
program's connection to the real world. The ensemble coordinates its activities
by cooperation rather than overall management. In the heterarchy, some programs call others in sequence, temporary managers of some process, but at each
level, every program passes execution to another when its particular job is

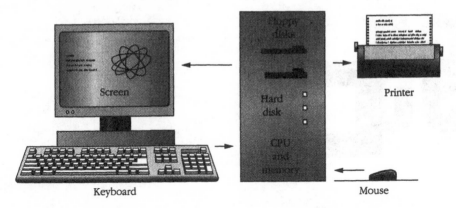

Figure 53.1 The operating system coordinates processes

done. Coordinating all of these processes is yet another function of the operating system.

One can begin to appreciate the rule played by a disk operating system by following the course of events when a user turns on a microcomputer. The very act of switching on the power initiates the *booting sequence*. The word "boot" expresses the amusing metaphor of a bootstrap, as when someone "lifts themselves by their own bootstraps"—a process that seems doomed to failure.

Every microcomputer comes equipped with special memory that contains certain key programs and data. The memory, called *read-only memory* or ROM, can be read but not written. When the microcomputer's power comes on, the programs stored in the ROM are activated as if they were programs in normal memory. Powering up the computer causes the command in a special built-in location to be executed. This is the first command in the bootstrap program. It initializes the hardware, making sure that all logical elements are in the correct state and it sets up the input and output facilities. It also reads a special location of the hard disk where it finds the bootstrap loader program. It immediately brings the bootstrap loader into main memory and then transfers execution to it. Early microcomputers, having no hard disks, had to read the bootstrap loader from transportable or floppy disks. Consequently, most disks had (and continue to have) a boot *sector* where the bootstrap loader resides.

Once in memory, the bootstrap loader consults another special area of the hard disk that contains details on the layout of the disk filing system. Here it locates two major portions of the operating system, a file manager and an I/O specialist. The bootstrap loader than transfers these and other portions of the operating system into main memory.

The file manager maintains logical information about files; their names, lengths, types, and locations on the disk. The I/O specialist manages low-level aspects of disk access. Thus it reads the file data into memory, checks for errors, recovers from them, moves the read/write head of the disk, and so on. Together, the file manager and the I/O specialist work in concert with the operating system to manage files for the user.

A disk operating system consists of four main parts, three of which have already been mentioned: the disk bootstrap, the I/O specialist, the file manager, and the command processor.

From the point of view of a computer user, the DOS provides a number of direct services. At the level of the DOS, before an application program is run, the user may call the command processor to format or check a floppy disk, for example. The DOS also enables the user to change the way the printer functions or the way output is displayed on the screen. The DOS may contain a batching feature which enables the user to write and store short customized sequences of DOS commands for particular jobs.

From the programmer's point of view, the DOS offers almost complete control of the way the computer operates. The programmer may employ DOS utility programs for controlling and/or redirecting input from the keyboard or mouse and output to the screen and printer. The programmer also has direct access to the DOS file manager and may use its commands to create and delete files, to access particular parts of the hard disk or floppy disk. Most important, perhaps, are the software tools that the DOS offers a programmer: a compiler for translating high-level languages into assembly language, an assembler for reducing the instructions of assembly language programs into machine language, libraries to support programming, a linker to combine several programs into a single memory load, and a debugger to aid in the elimination of execution errors.

Most disk operating systems offer a simple editor for creating and changing files. Called a line editor, it works one line at a time. As well as any single feature of the DOS, it illustrates the system in action (Figure 53.2). What happens when someone types in the command that calls up the line editor?

edit message.txt

As this command is typed out at the keyboard, each character sends a signal representing itself to device-handling hardware. Called interrupts, the signals are stored in a special register until the interrupt generated by the carriage return arrives. At such a point, the device handler passes the entire string

edit message.txt

to a command interpreter, part of the command processor. The command   **353**

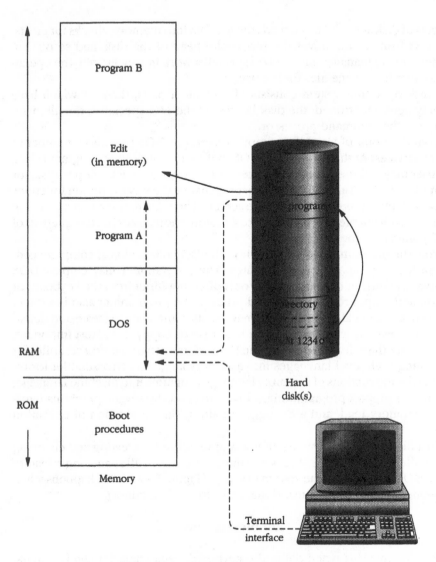

Figure 53.2 The DOS locates the Edit program

interpreter gives priority to the word **edit.** It saves the rest of the string **message.txt** in main memory, then uses a special program called a system loader to search for a program by the name of **edit.**

The DOS maintains special systems directories both on its hard disk and in main memory. Each directory contains the starting address and file size for every program and file stored there. If the sought-for program or file is not in the main memory, the file editor searches appropriate areas of the hard disk to see if **edit** is there. To actually carry out the search, the system loader must interact with the file manager and I/O specialist (not to mention other parts of the DOS such as the memory manager) to read the file data for the disk and to locate **edit** properly in main memory. Once it locates the directory entry for **edit,** the command interpreter knows where the program is and how much memory it will require.

Next in the chain of events leading to the appearance of **edit,** the memory manager comes into play. It sets aside a section of main memory for the **edit** program to live in while it works, so to speak. The system loader than loads **edit** into the reserved section, storing the starting and ending addresses in the memory manager's address map where it keeps all the real and relative addresses and their cross-references.

When the **edit** program is ready to run, the memory manager creates a process that will monitor the operation of **edit** on behalf of the user, starting the program up and ensuring successful termination when the editing session is complete.

To sum it all up, once the DOS is booted, it controls the entire operation of the computer. Driven by the commands in the command buffer, it loads programs specified by the user, monitors them as they run, and seizes control back when they terminate. The DOS also manages all disk and terminal traffic. A disk operating system resembles an air traffic controller. Writing a complete operating system is probably the most complex programming project one can imagine.

Problems

1. Suppose a disk operating system has a command

$$\text{read disk } track, locn$$

which reads the entire disk track specified by *track* into memory locations *locn* to *locn*+*n*−*1*, where *n* is the size of the track. Design a ROM program to perform the bootstrap load of the DOS. Assume that the computer starts at location 0 when the machine is powered up. Be sure to indicate where ROM ends and RAM begins.

2. Assuming that a disk has 1000 tracks, each of which can hold 128 datablocks of 1020 characters of information, design a directory structure for this disk. Assume that the file names have less than 14 characters.

3. For a CPU to "overlap I/O" means that it will continue regular processing while data is being transferred. Suppose a computer comes equipped with a special process that, when a certain interrupt is received, will store in location 0 the location of the instruction currently being executed. The process then executes the instruction in location 1000. How would you program this computer to read in as much data as it can and still keep the CPU as busy as possible?

References

Milan Milenkovic. *Operating Systems — Concepts and Design.* McGraw-Hill, New York, 1992.

Richard Allen King. *The MS-DOS Handbook.* Sybex, Berkeley, Calif., 1985.

54

NP-COMPLETE PROBLEMS

The Tree of Intractability

At present over 2000 problems are known to be NP-complete (see Chapter 41). Since 1972, when Stephen Cook found the satisfiability problem to be NP-complete, researchers have published a steady stream of NP-completeness results for other problems. The list of NP-complete problems is continually growing.

The problems on this list come from a wide variety of sources — computer science, engineering, operations research, mathematics. Some of the problems arise from practical applications such as compiler optimization, analysis of structural steel frameworks, and shop scheduling. Others arise in a purely theoretical context such as the theory of equations, calculus, or mathematical logic.

A problem is shown to be NP-complete by finding a special transformation to it from a problem already known to be NP-complete. A tree of such transformations connects SAT, the satisfiability problem to all NP-complete problems (Figure 54.1). Such problems include the vertex-cover (VC) problem, the traveling salesperson problem (TSP), and many others. The transformation which shows that a particular problem is NP-complete shares two important properties with Cook's original generic transformation (see Chapter 45):

It can be completed in polynomial time.
It preserves solutions.

Suppose, for example, that A and B are two problems in NP and that f is a transformation from A to B with these properties. If A is NP-complete, there is a generic transformation g from every problem in NP to problem A. It does no harm to think of this generic transformation as the one specified in Cook's theorem. In other words, for each problem X in NP there is a polynomial q such that each instance k of X is transformed to $g(x)$ (an instance of A) in time $q(n)$, where n is the size of x. But we also have a transformation f that acts on $g(x)$ to produce the instance $f(g(x))$ of B in time $p(m)$, where m is the size of $g(k)$. It takes little further argument to establish that the composite transformation $f(g)$ takes x into an instance of B in no more than $p(q(n)) + q(n)$ steps. It is also true that x is a yes instance if and only if $f(g(x))$ is a yes instance.

These are the essential elements in a proof that problem B is NP-complete. When one is faced with a brand-new problem suspected of being NP-complete, the usual first step is to show that it lies in NP: Ensure that it can be solved in polynomial time by a nondeterministic Turing machine (or computer). The next step is to find a specific transformation like f above from a problem already known to be NP-complete to the new problem.

A pretty example of an NP-completeness transformation involves satisfiability and the problem of covering all the edges of a graph by a specified number of vertices. Called the *vertex-cover problem* (or VC for short), the question is whether all the edges of the graph can be covered by a given number k of

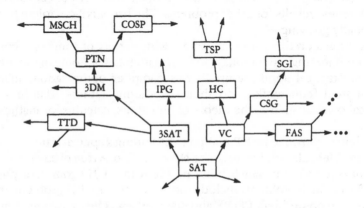

Figure 54.1 The NP-completeness tree

vertices; each edge must be incident with at least one of the k vertices. One way to settle the question is to find a subset V of vertices such that

Every edge in G has at least one of its vertices in V
V has k (or fewer) vertices.

This problem lies in NP (see Problem 1 in Chapter 41).

The transformation f from SAT to VC can be described entirely in terms of its effect on the clauses of an instance of SAT. First, let the variables appearing in a specific instance x of SAT be listed as

$$x_1, x_2, \ldots, x_n$$

Replace each variable x_i by an edge (u_i, v_i) joining two vertices u_i and v_i in a graph that is constructed as we go. When this process is complete, the graph has n edges and $2n$ vertices.

The next step is to replace the ith clause

$$(z_1, z_2, \ldots, z_m)$$

by a complete subgraph on m vertices $w_{i1}, w_{i2}, \ldots, w_{im}$. Each z_j is a literal, namely, a specific boolean variable or its negation. For simplicity's sake, assume that the variables are $x_1, x_2, \ldots, x_m$ $(m \leq n)$.

The next step in constructing the VC instance output by f involves specifying how the vertices of the complete subgraphs are joined to the u and v vertices. Briefly, if $z_j = x_l$, then join w_{ij} to u_l. But if $z_j = \sim x_l$, join w_{ij} to v_l.

The final step is to specify the integer k: Let k be the number of variables plus the number of occurrences of literals minus the number of clauses.

The construction carried out by f is illustrated in Figure 54.2 for the following SAT instance:

$$(x_1, x_2, \sim x_3)(\sim x_1, x_3)(x_1, \sim x_2, x_3)(x_2)$$

Can there be any doubt that f is computable in polynomial time? First, f does not really draw a graph like the one in Figure 54.2. It merely needs to create a list of its edges. The replacement operations are straightforward in most cases. The construction of a complete subgraph can be a bit time-consuming, however, since its size goes up as the square of the number of literals in the clause it represents. But this is still a polynomial size and can be handled by a polynomial-time program, whether for a Turing machine or a random access machine.

As for preserving "yes-ness" of instances, the argument requires some spelling out. As a general rule, there are always two arguments to give. First, prove

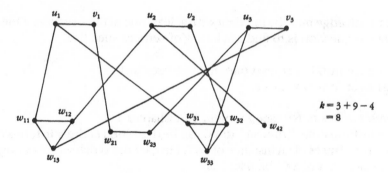

$$k = 3 + 9 - 4$$
$$= 8$$

Figure 54.2 The transformation from SAT to VC

that if *S* is a yes instance of satisfiability (or the NP-complete problem one started with), then *f*(*S*) is a yes instance of VC. Then prove the converse. If *f*(*S*) is a yes instance, so is *S*.

Suppose that *S* is a yes instance of satisfiability. Then *S* has an assignment of truth values to its variables that satisfies each clause. For each variable x_i made true by the assignment, let u_i be in cover *V*. If x_i is made false, on the other hand, place v_i in *V*. Since each clause contains at least one true literal, it follows immediately that each of the complete subgraphs constructed by *f* has at least one vertex, say *w*, that is adjacent to a vertex in *V*. Do not put *w* in *V*. Instead, add all the other vertices of that complete subgraph to *V*. It is not hard to see that the cover *V* ends up with no more than *k* vertices as specified in terms of the numbers of variables, literals, and clauses.

The converse argument is now also clear. Given a cover *V* for the graph *G* constructed by *f*, the integer *k* bounds the size of *V*. However, each of the subgraphs (u_i, v_i) will require at least one vertex in *V* to be covered. Each of the complete subgraphs has at most one vertex *not* in *V*. The minimum number of vertices required by this reckoning is the number of variables plus the number of literals minus the number of clauses—precisely *k*. The "tightness" of the construction thus ensures that each of the subgraphs mentioned has precisely the minimum number of covering vertices allowed: Each subgraph (u_i, v_i) has one, and each complete subgraph on *m* vertices has *m* − 1 vertices in *V*.

For each u_i in *V*, assign x_i = true. For each v_i in *V*, assign x_i = false. As already observed, each subgraph contains a *w* vertex not in *V*. But that vertex must be adjacent either to a *u* or *v* vertex. In either case, the latter vertex must be in *V*, and the literal corresponding to the *w* vertex must be true. Thus every clause contains at least one true literal under the assignment. The SAT instance is satisfiable. The answer is yes.

This completes the sample NP-completeness transformation. Of the many proofs of NP-completeness now in the literature, a good percentage are no more complicated than this one. Others, however, are much more complicated. The subject is almost a separate area of computer science all by itself.

The existence of so many NP-complete problems has resulted in a certain attitude of pessimism. Each new NP-complete problem merely reinforces the idea of how hopeless it is to hunt for a polynomial-time algorithm for an NP-complete problem. Cook's generic transformation maps all the problems in NP, particularly those that happen to be NP-complete, into SAT. But SAT is equally transformable to any NP-complete problem. This observation boils down to realizing that any two NP-complete problems are mutually transformable in polynomial time. Solve one, and you have solved them all. But no one has ever found a polynomial-time algorithm for any NP-complete problem.

Although it is not of immediate practical importance to solve decision problems like the ones in NP, it is certainly important to solve various closely related problems that have more complex answers. For example, for a given graph, one might wish to know the minimum number of vertices in a covering set. One would probably want an algorithm to output such a set. But if, by merely asking whether a graph has a cover of a certain size, one is already asking a question that apparently cannot be settled in polynomial time, then the original problem is at least as tough. Such problems, ones that require more elaborate solutions than yes or no, and called *NP-hard*.

Faced with the need for solving NP-hard problems in polynomial time, computer scientists have been driven increasingly to searching for approximate algorithms that run in polynomial time.

Problems

1. The three-satisfiability problem (3SAT) is almost the same as the satisfiability problem. The only difference is that 3SAT requires each clause to contain exactly three literals. Show that 3SAT is NP-complete by finding a polynomial-time transformation from SAT. The heart of this transformation is to replace an arbitrary clause by a logically equivalent set of 3-clauses. Use additional logical variables, if necessary.

2. Show that any two NP-complete problems are polynomial-time equivalent – that there are polynomial-time transformations running in both directions between the problems.

3. If $P \neq NP$, explain why there can be no polynomial-time transformation from *VC* to graph matching (GM). In this problem one is given a graph *G* and an integer k: Are there k or more edges in G such that no two share a common vertex?

(*Hint:* An $O(n^{2.5})$-time algorithm is known for GM.)

References

Alfred V. Aho, John E. Hopcroft, and Jeffrey D. Ullman. *The Design and Analysis of Computer Algorithms.* Addison-Wesley, Reading, Mass., 1974.

M. R. Garey and D. S. Johnson. *Computers and Intractability: A Guide to the Theory of NP Completeness.* Freeman, San Francisco, 1979.

55

ITERATION AND RECURSION

The Towers of Hanoi

The famous towers of Hanoi puzzle (Figure 55.1) involves three pegs and n disks of diameter 1, 2, 3, . . . , n. The puzzle begins with all the disks stacked on one of the pegs in order of increasing size from top to bottom. One is asked to transfer all the disks to one of the other, pegs obeying the following rules:

1. Move only one disk at a time.
2. Never place a disk on top of a smaller one.

The origin of this puzzle is attributed (perhaps with some reservations) by W. Rouse-Ball, the great English writer on mathematical recreations, to the following legend.

In the great temple at Benares . . . beneath the dome which marks the center of the world, rests a brass plate in which are fixed three diamond needles, each a cubit high and as thick as the body of a bee. On one of these needles, at the creation, God placed 64 disks of pure gold, the largest disk resting on the brass plate, and the others getting smaller and smaller up to the top one. This is the Tower of Bramah. Day and night unceasingly the priests transfer the disks from one diamond needle to the other. . . . When the 64 disks shall have been thus transferred [i.e., according to rules 1 and 2 above] from the needle on which at the creation God placed them to one of the other needles, then tower, temple, and Brahmins alike will crumble into dust, and with a thunderclap the world will vanish.

The originator of this tradition clearly expected the priests to take a long time to transfer the disks. Indeed, as the reader may already know, the transfer of n disks cannot be completed in fewer than $2^n - 1$ moves. This means that even if the priests could make a move once every second, the transfer would still require nearly 600 trillion years, a figure which dwarfs estimates by modern cosmologists of the probable lifetime of the universe.

In any event, the algorithm for solving the towers of Hanoi puzzle is usually presented in recursive form and has often been cited in textbooks as an example of recursion.

HANOI

procedure: *hanoi* (n, A, B)
    if $n = 1$
    then move disk from A to B
    else *hanoi* $(n - 1, A, C)$
        move disk from A to B
        hanoi $(n - 1, C, B)$

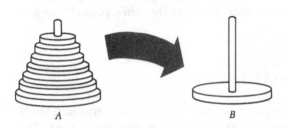
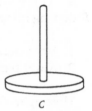

Figure 55.1 Move all the disks from peg A to peg B

Here, the procedure *hanoi* moves *n* disks from peg *A* to peg *B* by first moving $n - 1$ disks from *A* to *C*, moving the remaining disk from *A* to *B*, and then moving the $n - 1$ disks on *C* back to *B*, completing the solution.

This neat, easily understood algorithm would certainly have the greatest impact if the problem could be solved in no other way, that is, if the only way to get all the disks from peg *A* to peg *B* in $2^n - 1$ moves were through some (presumably similar) recursive procedure. In fact, there is a purely iterative solution due to Peter Buneman and Leon Levy who published it in 1980 (see Figure 55.2). Here *A, B* and *C* are arranged in clockwise order.

The Buneman-Levy algorithm is based on the observation that at any stage of a solution the smallest disk is available to move and can go to either of the remaining pegs. Since the smallest disk has been moved, it would be pointless to move it again, and there is only one other move possible at this stage: Move the second smallest exposed disk (let us just call this the *second* disk) to the other peg not occupied by the smallest disk. By restricting the smallest disk always to move clockwise, the following algorithm is immediate.

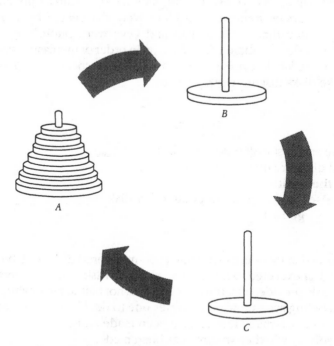

Figure 55.2 The Buneman-Levy algorithm operates in clockwise fashion

TOWER

procedure: *tower*
   repeat
       move smallest disk clockwise
       move second disk to remaining peg
   until all disks are on one peg

A proof that this algorithm solves the towers of Hanoi problem in $2^n - 1$ moves is based on the following theorem of Buneman and Levy:

> For any k between 1 and n, the procedure tower, in 2^k moves, transfers the k smallest disks (and only those disks) one peg clockwise if k is odd and two pegs clockwise if k is even.

A rather useful feature of the Buneman-Levy algorithm is the greatly reduced demands it makes on computer memory. For example, the recursive program running *hanoi* requires that up to n copies of the problem be stored at any one time (see Chapter 24). *Tower,* on the other hand, requires only one copy.

Somewhat fancier iterative algorithms were discovered by T. R. Walsh at almost the same time as Buneman and Levy were publishing. In Walsh's scheme, the disks are labeled 1, 2, . . . , n in order of increasing size, and pegs A, B, and C are labeled $n + 1$, $n + 3$, and $n + 2$, respectively. These pegs are treated like disks that are never moved:

WALSH

1. move smallest disk onto even disk
2. **if** all disks are on the same peg
      then exit
      else move second disk onto other disk
         go to 1

Walsh shows that there is exactly one even-numbered disk available each time statement 1 is executed, whence the "second" disk is moved onto the other available disk, provided that the algorithm has not halted meanwhile. Walsh has a binary labeling scheme which enables one to decide, on examining a given configuration, how many moves have been made and (in the case of a human problem solver) whether an error has been made.

Elsewhere we have examined the ideas underlying recursion and have displayed some elegant recursive programs. Indeed, the idea of recursion as applied to computer programs has been one of the most powerful and fruitful themes in computer science. The "home ground" of recursion is undoubtedly its expressive power and simplicity. However, recursive algorithms are not always the most efficient in their use of either time or space resources, and this is an observation which both programmers and theorists would do well to keep in mind.

Problems

1. Show that the n-disk Towers of Hanoi problem cannot be solved in fewer than $2^n - 1$ moves. the simplest way of doing this is to establish the result for 1 disk, assume it for $n - 1$ disks, and then to prove it for n disks by an inductive argument.

2. *Any* recursive algorithm can be rewritten as an iterative algorithm by using a pushdown stack to keep track of information relating to each call. However, any iterative algorithm can be rewritten as a recursive one. Do this for the Buneman-Levy algorithm, using the kind of algorithmic notation developed in this book.

References

Leon S. Levy. *Discrete Structures of Computer Science*. Wiley, New York, 1980.

P. Berlioux and P. Bizard. *Algorithms: The Construction, Proof, and Analysis of Programs* (trans. Annwyl Williams). John Wiley & Sons, New York, 1986.

56

VLSI COMPUTERS

Circuits in Silicon

In the last 20 years, computer hardware has undergone a veritable revolution. From printed-circuit boards studded with transistors, capacitors, and resistors, basic computer functions have migrated to a new technology of ultrasmall wires and components etched on the surface of silicon chips. The latter technology is called *very large scale integration* (VLSI).

In other chapters devoted to computer logic in this book, almost no attention has.been paid to the underlying physical hardware. The reasons for this approach lie in the universality of logic: No matter how logic is implemented, basic functions from AND to NAND are involved (see Chapter 3). A circuit that is designed for one set of gates is relatively easy to redesign with other gates as basic elements. Logic principles remain the same.

VLSI circuits are fabricated on a three-layer substrate (Figure 56.1). The uppermost layer consists of a metal such as aluminum, and the middle layer consists of polysilicon. The bottom layer is a single crystal of nearly pure silicon that is "doped" with certain impurities to produce desirable electrical properties. It is called the *diffusion layer*. Between each pair of adjacent layers is a thin, insulating layer of silicon dioxide. The layers are built up, one at a time, in

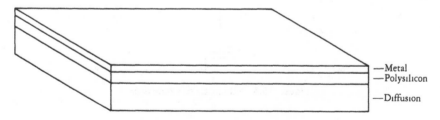

Figure 56.1 The physical substrate

patterns that reflect the circuit being constructed. Typically, the patterns are built up of narrow strips of material called *paths.*

Whenever a path in the polysilicon layer crosses a path in the diffusion layer, the voltage in the first path controls the voltage in the second; a transistor results. Shown below is a magnified portion of two such paths separated by a layer of insulating silicon dioxide (Figure 56.2). Layers of silicon chip technology can be built up in patterns either by adding material selectively or by removing it selectively. In the lowest level of the magnified fragment above, the diffusion layer has been neither added to nor subtracted from. Instead, it has been doped by diffusing a special element like phosphorus or boron. In the first case, pure silicon that is doped with phosphorus ends up with extra electrons that are free to act as carriers of negative electric charge within the diffusion pathway. In the second case, doping with boron removes electrons from the silicon, creating "holes," or carriers of positive charge.

Fabrication processes such as masking and etching have left a strip of silicon dioxide supporting a strip of polysilicon. The ability of voltage in the polysilicon path to control voltage in the diffusion path yields a basic transistor, ren-

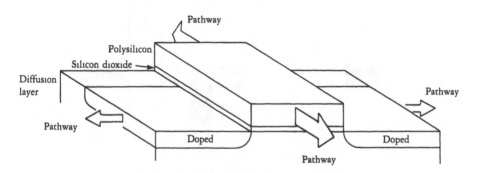

Figure 56.2 Two pathways form a transistor

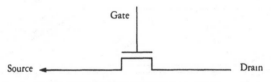

Figure 56.3   Schematic of a transistor

dered schematically in Figure 56.3. The amount of charge, or voltage, in the gate (polysilicon pathway) controls voltage in the diffusion pathway as follows: If this charge exceeds a certain threshold level, electrons will accumulate in the undoped region of the diffusion layer just below the gate. These excess electrons form a conducting pathway between the drain and the source. Consequently, with a below-threshold charge on the gate, there is virtually no current flowing from drain to source and thus there is a high voltage. When the gate charge rises over the threshold, current flows and the voltage drops. The result is a basic switch from which a number of logic components can be constructed.

For example, an inverter (see Chapter 13) can be constructed. An inverter converts one kind of signal to its opposite. In what follows, *high* will denote an above-threshold voltage, and *low* will mean a voltage that is below threshold. Seen from above, the various pathways making up the inverter might appear as in Figure 56.4.

Here, polysilicon layer regions are indicated by light gray, and diffusion layer pathways by dark gray. Below either polysilicon region there is no doping. Two transistors make up the basic inverter circuit. The polysilicon region on the left is connected to a constant voltage supply different from the supply pathway in the diffusion layer below it. The other transistor operates like the one described

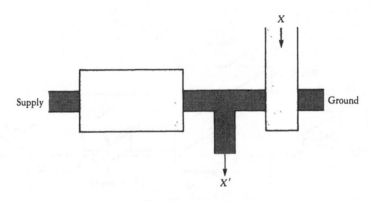

Figure 56.4   An inverter in silicon

earlier. The voltage entering its gate constitutes the input signal X. When this signal is high, the pathway labeled X' drops to the ground level, thus becoming low. But when the voltage in the gate labeled X is low, there is no connection between the pathway labeled X' and ground. At this point, the transistor on the left comes into play. Since its gate is always turned on (or high), there is always a current flowing from the supply pathway under the gate and into the pathway labeled X'. Instead of flowing directly to ground, this current now creates a charge in the pathway labeled X'; the new output signal is high. The basic transistor geometry just introduced can now be extended to two or more transistors of the righthand type. The resultant VLSI pattern (Figure 56.5) is only slightly more complicated than the previous one. In this circuit, the only way pathway $(XY)'$ can carry a low occurs when both gates X and Y carry a high. In this mode, the supply pathway is connected directly to ground via all three gates. Consequently, the voltage in pathway $(XY)'$, being also connected to ground, is low. In logic symbolism,

$$(XY)' = 0 \quad \Leftrightarrow \quad X = 1 \text{ and } Y = 1$$

But this is merely a definition of the logic function $(XY)'$; the pathway was correctly labeled, and the circuit operates as specified, namely, as a NAND gate.

Chapter 3 made it clear that any logic function can be realized by a combination of inverters and NAND gates. Thus, armed with only the transistor configurations introduced so far, one could fabricate all the elements of a digital computer. However, since NOR gates and other logic components are also available, the computer on a single silicon chip makes use of them whenever a designer finds it convenient.

So far, only the lower conducting layers (diffusion and polysilicon) have been discussed. The upper conducting layer is metallic. It makes connections

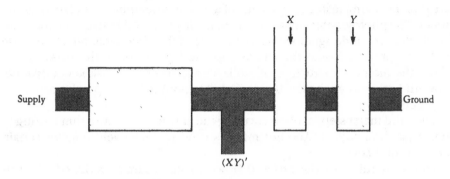

Figure 56.5 A NAND gate in silicon

371

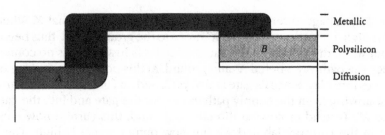

Figure 56.6

between the output pathways of one logic component and the input pathways of others. The connections are made through one or both insulating layers by means of holes created during the fabrication process. In schematic terms, the cross section of such a connection reveals that the metallic layer is not strictly confined to one level of the chip. Indeed this is also true of the polysilicon layer and the intervening insulating layers (Figure 56.6). Here the output of one logic device is represented by the diffusion pathway labeled *A*. This pathway is connected via direct metallic contact to the polysilicon pathway labeled *B*. The latter pathway might represent the input channel of another logic device.

The fabrication techniques that create the inner logic of a VLSI circuit have more in common with printing and lithography than with traditional electronics. From a single, large crystal of silicon grown under the most exacting conditions, a number of wafers are cut. Each wafer is several centimeters across and only millimeters thick. VLSI techniques may generate several hundred copies of the computing circuit (under production) on the surface of the wafer. Each circuit may be only a few millimeters wide, yet it may contain several thousand logic components!

The second major step in circuit fabrication creates a layer of silicon dioxide on the wafer surface. This is immediately followed by the selective removal of areas of insulating material by means of a tiny photographic mask laid over the wafer. The photographic image has been greatly reduced in size from the scale at which a circuit designer works to the scale of the microcircuit itself. Intense radiation projected onto the silicon dioxide layer through the mask breaks down the molecular structure of the layer wherever the mask is not opaque. Powerful solvents wash away the degraded regions to leave a precise pattern of exposed silicon.

The third major step involves covering the entire wafer with a thin coating of polycrystalline silicon. Another mask is used to selectively remove certain portions of this layer.

Following this step, the exposed (lower) silicon surface is doped with various impurities to create the sources and drains of all the transistors as well as to create additional conducting pathways between components. A second insulating layer of silicon dioxide is now laid over the surface in a pattern that ensures noncontact between the metallic layer to follow and either the polysilicon or the diffusion layer. Of course, wherever the metallic layer must make contact with either surface, the silicon dioxide layer is absent.

The last major stage of fabrication coats the entire wafer with a thin covering of metal such as aluminum. A final mask is used to remove all the metal except in areas where conducting pathways are desired. When the entire VLSI wafer is thus completed, it is cut into individual circuits. Each circuit is then cemented into a plastic package. Tiny wires connect certain metallic pathways at the edge of the chip to the pins that sprout from the package. The latter object is familiar to many of us as the mysterious little black box that resides in the heart of the computer.

Problems

1. Design a configuration of transistors (involving the diffusion and polysilicon layers alone) that functions as a two-input NOR gate.

2. Draw a series of masks that might be used to fabricate the NAND gate displayed in this chapter. Assume separate metallic pathways that contact each of the three gates, and use any geometry that seems appropriate, bearing in mind that the geometry in actual VLSI circuits is far from arbitrary.

References

Carver Mead and Lynn Conway. *Introduction to VLSI Systems.* Addison-Wesley, Reading, Mass., 1980.

Jeffrey D. Ullman. *Computational Aspects of VLSI.* Computer Science Press, Rockville, Md., 1984.

57

LINEAR PROGRAMMING

The Simplex Method

I magine a plane passing through a polyhedral, prismatic-looking solid (Figure 57.1). If the plane is moved upward to a final resting position (shown in dotted lines), it passes through exactly one point of the solid: it touches it. Who would think, with no prior information on the subject, that the polyhedral solid, the plane, and the point might represent an important problem in economics, scheduling, chemical process control, or the shipment of goods?

The polyhedron represents a set of linear inequalities which express constraints inherent in a given problem. The plane, in fact the position of the plane, represents the value of a certain linear function someone is trying to optimize. The point represents the optimal value of that function vis-à-vis the given constraints. With respect to the plane's orientation, the point is an exremal vertex of the solid.

In Figure 57.1, a path (in bold lines) is shown leading to the optimal point. This path represents the course traveled by the best-known algorithm for solving linear programming problems, the simplex algorithm. A set of linear in-

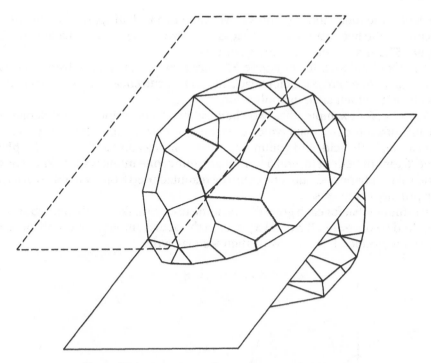

Figure 57.1 A geometry of constraints resembles a jewel

equalities, which we may write as

$$a_{i1}x_1 + a_{i2}x_2 + \cdots + a_{in}x_n \leq b_i \qquad \text{for } i = 1, 2, \ldots, m$$

represents m half spaces, each n-dimensional and each bounded on one side by a plane. The equation of this plane is obtained by replacing the inequality above by an equality. The half space represented by the ith inequality above consists of all points lying on one side of the plane. The intersection of all m half spaces, if nonempty, forms a convex, polyhedral solid, and each of its facets is contributed by one of the planes.

Consider the following two-dimensional example:

$$x_1 + 5x_2 \leq 8$$

$$2x_1 + 3x_2 \leq 6$$

$$3x_1 + x_2 \leq 6$$

$$-x_1 \leq 0$$

$$-x_2 \leq 0$$

Each of these inequalities defines a two-dimensional half space, and the intersection of the five resulting half spaces is shown as the convex shaded area in Figure 57.2. Its "facets" are line segments.

The set of inequalities in Figure 57.2 has a somewhat special form; the last two contain one zero coefficient each. Such an occurrence is not unusual in the inequalities which arise in applications.

Suppose, for example, that two chemical products, uniphane and duozene, are manufactured in a plant which uses three primary chemicals in the process: alphatone, betic acid, and gammine. If the plant has on hand 8 tons of alphatone, 6 tons of betic acid, and 6 tons of gammine, one might want to know how much of uniphane and duozene could be manufactured from available stocks of the primary chemicals.

It turns out that for each gram of alphatone used to make uniphane, 5 grams is required by duozene. If x_1 and x_2 denote the actual amounts of uniphane and duozene produced, respectively, then we may write

$$x_1 + 5x_2 \leq 8$$

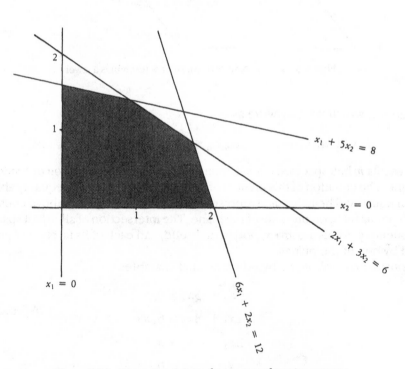

Figure 57.2 Constraints arising from a manufacturing process

the first inequality in the set shown above. Similarly, the next two inequalities arise from the relative proportions of betic acid and gammine used in the manufacturing process. The last two inequalities simply express the fact that nonnegative amounts of both uniphane and duozene must ultimately be produced.

Each point within the convex shaded region of the figure opposite represents a feasible combination of amounts x_1 and x_2 produced. The amounts, naturally, are expressed in tons, and any point outside the feasible region (the convex, shaded area) represents a pair of amounts which simply could not be produced from available chemical stocks.

The chemical company, however, is interested in not only how much uniphane and duozene *could* be produced but also how much *should* be produced. At this point, market forces come into play, and the market value of the two products is used to assign a value to each pair of production amounts (x_1, x_2) in the region of feasibility. If uniphane sells for $2 million per ton while duozene is worth $1 million per ton, then the value of amounts x_1 and x_2 is z, where

$$z = 2x_1 + x_2$$

For each possible value of z, this cost function may be represented by a straight line in our figure. All feasible points on this line represent production amounts x_1 and x_2, the combined market value of which is z. The chemical company wants to maximize z and thereby to maximize its profit. As the line is moved farther and farther from the origin, z increases (Figure 57.3). Finally, if it is moved any farther, it will no longer intersect the feasible region; in such a case, the corresponding z value would not be attainable by the company with the chemicals on hand. The best it can do is represented by the extremal point at which the cost line comes to rest, so to speak.

What is the best z value that the chemical plant can achieve? This depends on the coordinates of the extremal point in question. The point is obtained simply by solving a pair of linear equations

$$2x_1 + 3x_2 = 6$$
$$3x_1 + x_2 = 6$$

The solution is $x_1 = {}^{12}\!/_7$ and $x_2 = {}^6\!/_7$. Thus the company must produce ${}^{12}\!/_7 = 1.71$ tons of uniphane and 0.86 tons of duozene. The value of this production will then be $z = 4.29 million, the very best it can do under the circumstances.

The previous solution was obtained largely by visual means: We *saw* which extremal point represented the maximum z value and merely found its coordinates by solving the equations of the lines meeting there. If the problem had

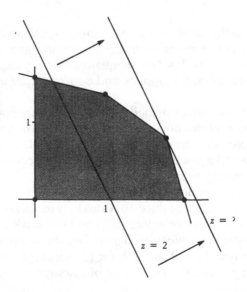

Figure 57.3 The most profitable levels of feasible production

much more than two variables, we would have been completely lost even trying to *conceive* of the intersecting plane producing the point, let alone actually visualizing it. Automatic methods are called for.

One such method, in constant use since its discovery in 1951 by George B. Danzig, is called the *simplex method.* It works not with inequalities of the above sort but with equations. To convert the foregoing system to equations, we introduce "slack variables" s_1, s_2, and s_3. These have no physical meaning as far as our chemical plant problem is concerned; they merely serve to convert inequalities to equalities by taking up the slack, so to speak. If we add the objective function to the equations thus obtained, we have the following system:

$$
\begin{aligned}
z - 2x_1 - x_2 &= 0 \\
x_1 + 5x_2 + s_1 &= 8 \\
2x_1 + 3x_2 + s_2 &= 6 \\
3x_1 + x_2 + s_3 &= 6
\end{aligned}
$$

The resulting system has four equations in six unknowns, and an initial solution is easily obtained by putting the system in diagonal form. This is normally done by the process of gaussian elimination, but in the case above, the system is almost already in this form:

$$-z \quad + 2x_1 + \ x_2 = 0$$
$$s_1 \quad + \ x_1 + 5x_2 = 8$$
$$s_2 \quad + 2x_1 + 3x_2 = 6$$
$$s_3 + 3x_1 + \ x_2 = 6$$

We obtain what is called an *initial feasible solution* from the equations in this form by setting $x_1 = x_2 = 0$.

The simplex method operates on systems of equations in this form, continually isolating one set of variables after another. The solutions so obtained represent points on the original convex feasible region, and each point yields a z value which improves the one obtained in the previous step—until the optimal value of z is reached. For the time being, we note merely that $z = 0$ in our initial solution is far short of the known optimal value of 4.29.

At each step in the simplex method, the coefficients of the objective equation are examined, and the largest is selected as the basis of the next step. This step, called a *pivot*, involves exhanging an isolated variable with a nonisolated one. Initially in our example the isolated, or *basic*, variables are z, s_1, s_2, and s_3. The new basic variable is x_1 because it has the largest coefficient in the objective function, namely, 2. To find out which basic variable x_1 will replace, we examine the ratios

$$\frac{a_{i1}}{b_i}$$

and select the equation with the largest such value. In our example, these ratios are ⅛, ⅔, and ⅜, respectively. The last ratio is the largest. It arises from the last equation, and so it is the basic variable s_3 which will be replaced by x_1.

The actual replacement, or "pivoting," is done by the process of gaussian elimination: Use the last equation to eliminate variable x_1 from all other equations by adding appropriate multiples of the last equation to the others. The new system is

$$-z \quad - \tfrac{2}{3} s_3 + \tfrac{1}{3} x_2 = -4$$
$$s_1 \quad - \tfrac{1}{3} s_3 + 14\tfrac{1}{3} x_2 = \ \ 6$$
$$s_2 \quad - \tfrac{2}{3} s_3 + \tfrac{7}{3} x_2 = \ \ 2$$
$$x_1 + \tfrac{1}{3} s_3 + \tfrac{1}{3} x_2 = \ \ 2$$

379

Again, a new basic solution is obtained by setting the nonbasic variables to 0. Thus, with $s_3 = x_2 = 0$, we obtain

$$z = 4 \qquad s_1 = 6 \qquad s_2 = 2 \qquad \text{and} \qquad x_1 = 2$$

Referring to the initial diagram of the chemical plant system, we notice that the values of x_1 and x_2 in this latest basic solution yield the extremal point $(2, 0)$ at which the objective function clearly has value 4. This is a great improvement over the previous value of 0, yet still a small distance from the absolute maximum of 4.29. At least one more pivot is required.

To select the pivot variable, we examine the coefficients in the objective function and notice that the largest belongs to variable x_2. The largest coefficient/constant ratio

$$\frac{a_{i2}}{b_i}$$

is $^{14}/_{18}$, obtained from the second equation. Thus the new basic variable is x_2 which replaces s_1, the basic variable inhabiting the second equation. When gaussian elimination is carried out a second time with this pivot, we obtain the final, optimal solution of $x_1 = {}^{12}/_7$ and $x_2 = {}^{6}/_7$ with the accompanying maximum z value of $^{30}/_7$ or roughly 4.29.

The simplex algorithm stops when there are no positive coefficients left in the objective equation. When the algorithm is implemented on a computer, the coefficients, constants, and basic variable list are maintained in arrays called *tableaux*.

The theory of linear programming goes well beyond the simplex method and its theoretical support. There are linear programs in which the objective function is to be minimized instead of maximized and in which the inequalities all have opposite directions.

In the last two decades, there has been much analysis of the simplex method. And it has been found, for example, that its worst-case performance, though very unlikely, can be very bad indeed: In some cases the number of pivots needed is exponential in the number of variables! By the time that some researchers had begun to wonder whether the linear programming problem was computationally intractable (see Chapter 41), a brilliant new algorithm was discovered by L. G. Kachian, an Armenian mathematician. Kachian's algorithm always runs in a polynomial number of steps. Briefly stated, it surrounds the region of feasibility by an n-dimensional ellipsoid that is systematically shrunk until it intersects the feasible region. Ultimately, it intersects the region at an extremal point.

Problems

1. Carry out the last gaussian elimination in the chemical plant example and thus obtain the optimal solution.

2. Write a complete algorithmic description of the simplex method for the kind of general problem described at the beginning of this chapter.

3. Sometimes a system of inequalities is so simple that no recourse to the simplex method is required. Given the inequalities

$$x_i \le b_i \qquad i = 1, 2, \ldots, n$$

and the objective function $z = c_1 x_1 + c_2 x_2 + \cdots + c_n x_n$, use your geometric intuition to discover a fast method for finding an optimal solution of the system.

References

T. C. Hu. *Integer Programming and Network Flows.* Addison-Wesley, Reading, Mass., 1970.

William H. Press, Brian P. Flannery, Saul A. Teukolsky, and William T. Vetterling. *Numerical Recipes: The Art of Scientific Computing.* Cambridge University Press, Cambridge, 1986.

58

PREDICATE CALCULUS

The Resolution Method

The predicate calculus is one of the most powerful languages known for the expression of mathematical ideas and thought. It has influenced computer science both directly through the invention of computer languages like PROLOG and indirectly through the theories of computation that depend on it.

As an example of the expressive power of the predicate calculus, consider the well-known problem of the wolf, the goat, and the cabbage (Figure 58.1). A man wishes to take a wolf, a goat, and a cabbage across a river. There is a rowboat he may use, but it is only big enough to hold one of the other living creatures besides himself. The problem is that he cannot leave the wolf and the goat together on one bank while he rows to the other: The wolf will eat the goat. Similarly, he cannot leave the goat and cabbage alone together.

Many possible configurations might be reached by the man, wolf, goat, and cabbage in the course of ferrying operations. These may all be represented by a single predicate P. It is really a truth function, and it has four arguments m, w, g, and c that take binary values. For example $m = 1$ will mean that the man is on the near side of the river, but $m = 0$ will mean that he is on the far side. A similar

Figure 58.1   A man ponders how to ferry a wolf, a goat, and a cabbage

convention will hold for w (the wolf), g (the goat), and c (the cabbage). The predicate is written functionally as

$$P(m, w, g, c)$$

and it has a very specific interpretation: For each possible combination of values of the four logical variables, P will be true for these values if and only if the corresponding configuration can be attained without the wolf eating the goat or the goat eating the cabbage. We will revisit the river presently.

The major interest by computer scientists in predicate calculus has been to exploit its expressive power to prove theorems. In the field of artificial intelligence, there have been many attempts to construct programs that could prove theorems automatically. Given a set of axioms and a technique for deriving new theorems from old theorems and axioms, would such a program be able to prove a particular theorem given to it? Early attempts faltered because there seemed to be no efficient technique for deriving new theorems. Then, in 1965, J. A. Robinson at Syracuse University discovered the technique called *resolution*. Not only does resolution frequently permit the derivation of theorems, but also it may be the basis for a new generation of computers currently under development. The so-called fifth-generation machines may spend all their time proving theorems, in effect. But such theorems will be disguised as problems in retrieval and deduction of facts from data bases (see Chapter 64). Even the possibility of the man successfully ferrying the wolf, goat, and cabbage across the river can be expressed as a theorem.

Syntactically the predicate calculus may be defined recursively as strings of symbols following a particular mode of construction — just as the propositional calculus was defined (see Chapter 13).

The basic building blocks of formulas in the predicate calculus are individual symbols. In the somewhat informal approach taken here, the uppercase alphabetic letters denote predicates while lowercase letters indicate functions and variables. It is understood that if more predicates, functions, or variables are needed than the roman alphabet can supply, then subscripting is used. Another liberty taken is to write integers in the normal way instead of in unary notation.

Besides the alphabetic symbols we will need parentheses; a collection of standard logic operators such as $\land$, $\lor$, $\rightarrow$, $\sim$; and two symbols peculiar to the predicate calculus. These are called the *existential* ($\exists$) and *universal* ($\forall$) *quantifiers,* respectively.

A *term* is defined recursively as follows:

1. A term is a variable.
2. If f is a function of n arguments and $x_1, x_2, \ldots, x_n$ are terms, then $f(x_1, x_2, \ldots, x_n)$ is a term.

An *atomic formula* is defined as any predicate the arguments of which are terms. Finally, the goal of these definitions, a *formula*, is defined recursively as follows:

1. An atomic formula is a formula.
2. If F and G are formulas, then so are $(F \lor G)$, $(F \land G)$, and $\sim (F)$.
3. If F is a formula and v is a variable, then $\forall v(F)$ and $\exists v(F)$ are formulas.

Of note in the foregoing definition is the use of symbols like x_1 for term and F for formula; these are not elements of the predicate calculus but rather part of a metalanguage used to define it. An example of a predicate calculus formula (hereafter simply called a *formula*) follows:

$$\forall e(P(e) \rightarrow (\exists d(P(d) \land (L(a(x,y),d) \rightarrow L(a(f(x),f(y)),e)))))$$

It is perhaps unfair to expect students of the theory of functions to recognize in this formula the definition of continuity of a real function. In the informal style of mathematics textbooks, the definition is normally written as follows:

$$\forall \epsilon > 0 \; \exists \delta > 0 \text{ such that } |x - y| < \delta \Rightarrow |f(x) - f(y)| < \epsilon$$

In the more restrictive setting of predicate calculus, the statement must be recast in terms of variables, functions, predicates, and their legal combinations into formulas. For example, $\forall \epsilon > 0$ is written $\forall e(P(e) \rightarrow \ldots)$. Here $P(e)$ is a predicate whose single argument is a real variable. It is interpreted to mean that e is a positive number. If e is positive, the formula goes on to say, there exists a d such that d is positive and such that the formula

$$L(a(x,y),d) \rightarrow L(a(f(x), f(y)),e)$$

is true. Here L is a predicate that is true when its first argument is less than its second. The function a defines the absolute value of the difference of its arguments. Under the interpretation just given for the formula as a whole, it obviously expresses continuity of a real function.

Of course, such a formula is not really equivalent to the definition of continuity until additional formulas defining the notion of real number, inequality, and so forth are given. Indeed, the more rigorous textbooks on real functions do something rather close to this. In any event, not only can the theory of functions be cast in the predicate calculus, but so also can virtually the whole of mathematics. This was the aim of *Principia Mathematica,* the ambitious logical codification of mathematics undertaken in 1921 by the British mathematicians Bertrand Russell and Alfred North Whitehead.

The predicate calculus inherits many of the rules of manipulation of the propositional calculus. For example, De Morgan's law states that

$$\sim (A \lor B)$$

is logically equivalent to

$$\sim A \land \sim B$$

where A and B are propositions. If we replace A and B by arbitrary predicates, the rule still holds. Additional rules of manipulation concern the universal and existential quantifiers. For example,

$$\sim (\forall x(P(x)))$$

is equivalent to

$$\exists x(\sim (P(x)))$$

The rule also works in reverse gear, so to speak, with

$$\sim (\exists x(P(x)))$$

equivalent to

$$\forall x(\sim(P(x)))$$

In formulas it is necessary to distinguish between "free" and "bound" variables. Bondage in this case has to do with being under the spell of a particular quantifier. More precisely, if F is a formula and x is a variable in F, then for both $\forall x(F)$ and $\exists x(F)$ the variable is *bound*. Any variable in F not so quantified is called *free*.

As in the propositional calculus, some predicate formulas are true, and others are not. The truth of a formula depends, however, on how its predicates and functions are interpreted. An *interpretation* of a formula involves a universe U and an interpretation function I that maps each n-place predicate of the formula into an n-ary relation on U. It also maps each n-place function symbol of the formula into an n-place function from U into itself.

If a formula F contains a free variable, in general it will be impossible under any interpretation of F to determine its truth. For example, under the usual interpretation of the less-than predicate L mentioned earlier, we cannot say whether the formula

$$\forall x\, L(x, y)$$

is true or not. Obviously, it would be if we inserted $\exists y$ right after $\forall x$. In any event, a formula containing no free variables is called *satisfiable* if it has at least one interpretation in which it is true. It is called *valid* if it is true under all possible interpretations.

To determine the validity of a formula is not an easy matter. Indeed, even to decide whether it is satisfiable is unsolvable: There is no effective procedure that will decide the satisfiability of arbitrary formulas.

There are procedures, however, for deriving formulas from other formulas. The "other" formulas may be called *axioms,* and the formula to be derived may be called a *theorem.* To derive a theorem from a set of axioms has been the business of mathematicians (whether they knew it or not) from the time of Euclid and even earlier.

For computer scientists interested in automatic theorem proving, the procedure of choice is the resolution method. The axioms one begins with here have a certain form called *clausal.* A *clause* is simply a disjunction of predicates or their negations. Any formula in the predicate calculus can be put in this form.

Consider two clauses $(\sim P(x) \vee Q(x, y) \vee R(y))$ and $(P(x) \vee \sim S(x))$. In one of these clauses the predicate $P(x)$ appears, and in the other clause its negation is seen. Provided that no other predicate shares this property in respect of the two clauses, we may replace them by the disjunction of all the literals taken together—with $P(x)$ and $\sim P(x)$ excluded:

$$(\sim P(x) \lor Q(x,y) \lor R(y)) \qquad (P(x) \lor \sim S(x))$$

$$(Q(x,y) \lor R(y) \lor \sim S(x))$$

Unfortunately, resolution is not quite this straightforward in some cases. For example, the predicate being eliminated may have a more complicated structure. Suppose that $P(a, f(x))$ appears in one clause and $P(x, y)$ appears in the other. To *unify* two such predicates means to find a substitution for the variables occurring in them that brings them into precisely the same form. The substitution must also be the most general possible effecting this purpose, rather like finding the smallest common denominator while adding fractions. Such a substitution for the example just given would be $x = a$ and $y = f(a)$. Carrying out the substitution (throughout both clauses being resolved) results in the appearance of

$$\sim P(a, f(a)) \qquad \text{and} \qquad P(a, f(a))$$

At this point resolution can be carried out as above.

Suppose now that F is a theorem to be proved and that $A_1, \ldots, A_n$ are all axioms. Formally speaking, F and $A_1, \ldots, A_n$ are all clausal formulas in the predicate calculus. Existential quantifiers have been removed by the process called *skolemization:* If $\exists x$ appears in a formula, replace all occurrences of x bound by this quantifier by a particular instance $x = a$ which makes the expression under the scope of the $\exists$ true. Each variable of each clause is otherwise understood to be universally quantified—the $\forall$ signs are omitted. The resolution method proceeds by first negating the theorem F to be proved and then adjoining it to the axioms. The resulting system can then be written

$$\sim F, A_1, A_2, \ldots, A_n$$

Roughly speaking, the method consists in resolving pairs of clauses within this system until a contradiction is reached. Both a predicate and its negation result from resolution. If this happens, F has been proved. If this cannot be made to happen, F cannot be proved from the axioms.

Recall the predicate $P(m, w, g, c)$. The universe in which this predicate is interpreted consists of a river, a man, a wolf, a goat, and a cabbage. The variables m, w, g, and c are binary-valued and refer to one side of the river or another: If $m = 0$, then the man is on the initial side of the river. If $m = 1$, he is on the other side. So it is with the wolf, goat, and cabbage. The predicate

$$P(0, 1, 1, 0)$$

is true if it is possible for the man and cabbage to be on this side of the river and the wolf and goat on the other. It would be unfortunate if this predicate were to be true.

In addition to the possibility predicate, it is useful to have an equality predicate $E(x, y)$ which is true if x and y have the same value. A function f makes it possible to distinguish one side of the river from another: $f(0) = 1$ and $f(1) = 0$. In general, we use f to define the first axioms, those governing equality:

1. $(E(x, x))$
2. $(\sim E(x, f(x)))$

The permissible ferrying operations yield four more axioms. Only the derivation of the first is explained.

If the man and the wolf are on the same side of the river, then it is possible for the man to row the wolf across the river provided that the cabbage and the goat are on opposite sides of the river. This statement can be written in terms of the foregoing function and predicates as follows:

$$P(x, x, y, z) \wedge \sim E(y, z) \rightarrow P(f(x), f(x), y, z)$$

Rewritten in clausal form, the statement becomes the next axiom:

3. $(\sim P(x, x, y, z) \vee E(y, z) \vee P(f(x), f(x), y, z))$

The remaining permissible ferrying operations follow in clausal form:

4. $(\sim P(x, y, x, z) \vee P(f(x), y, f(x), z))$
5. $(\sim P(x, y, z, x) \vee E(y, z) \vee P(f(x), y, z, f(x)))$
6. $(\sim P(x, y, x, y) \vee E(x, y) \vee P(f(x), y, ,x, y))$

Two more axioms complete the basis for a solution to the problem. The first represents the initial condition of the man, wolf, goat, and cabbage. The second represents the final condition. This clause is negated.

7. $(P(0, 0, 0, 0))$
8. $(\sim P(1, 1, 1, 1))$

Shown below is a set of unifications and resolutions laid out in the form of a diagram.

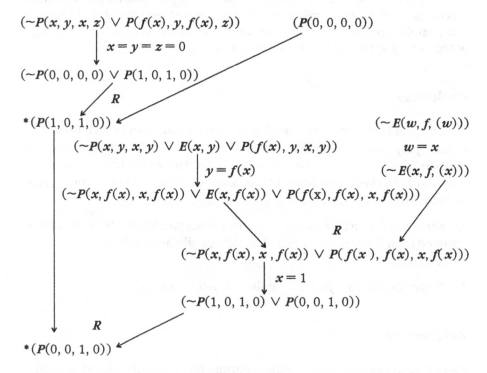

The two clauses marked with an asterisk represent two steps toward a solution of the ferrying problem. In the first, the man has taken the goat to the opposite side of the river. In the second, he has returned to the first side.

The resolution steps (marked R) were chosen with the solution in mind. A real resolution theorem-proving system would certainly not show such behavior. It would proceed rather blindly, looking for resolvents and unifying, building up a list of intermediate clauses until, finally, it would derive $(P(1, 1, 1, 1))$. Matching this against axiom 8, it would obtain a contradiction. This fact is guaranteed by a theorem of Robinson's which says that resolution arrives at a contradiction if and only if the formula originally negated is logically implied by all the axioms.

As a general rule, the resolution method is not implemented in helter-skelter fashion. There are numerous strategies that help the method coverage on a contradiction as quickly as possible. One is called the *unit preference strategy:* Choose those clauses (to unify and resolve) that are as short as possible. Another strategy, called the *set-of-support strategy*, maintains a distinction between primary axioms and supporting axioms; no two primary axioms are ever resolved against each other.

Resolution becomes rather unwieldy, from a computational point of view, for problems much more complicated than the ferrying problem. But deductions this complicated do not seem to be called for in the context of many logic programming applications (see Chapter 64).

Problems

1. The problem of the man, wolf, goat, and cabbage can be cast entirely in terms of propositions by taking the axioms as given and substituting all possible combinations of m, w, g, and c values into them. The set of clauses so obtained is called the *Herbrand universe*. In this case the Herbrand universe turns out to be finite. How big is the resulting propositional formula?

2. Show that if a set $\mathscr{S}$ of predicate clauses is not satisfiable, then neither is $\mathscr{S}'$ obtained from $\mathscr{S}$ by resolving two of its clauses, discarding them, and retaining the resolvent.

3. Finish the ferrying problem in the manner suggested.

References

Harry R. Lewis and Christos H. Papadimitriou. *Element of the Theory of Computation.* Prentice-Hall, Englewood Cliffs, N.J., 1981.

Alfred V. Aho and Jeffrey D. Ullman. *Foundations of Computer Science.* Chapter 14. Computer Science Press, New York, 1992.

59

THE HALTING PROBLEM

The Uncomputable

The Turing machine concept is one of several equivalent formulations of what we mean by "effective procedure" or "computation." Nothing more powerful than a Turing machine has been discovered that captures any better the meaning of such terms (see Chapter 17). And yet there are limits to the power of Turing machines, problems that Turing machines cannot solve. Such problems parallel those for which no effective procedure or recursive computation exists. In fact, one may reformulate a Turing-unsolvable problem as one which is unsolvable in any of the equivalent formal systems.

The best-known such problem is called the *halting problem*. It asks for a Turing machine T_H (Figure 59.1) which is able to perform the following task for any pair (T, t) as input.

"Given an arbitrary Turing machine T as input and an equally arbitrary tape t, decide whether T halts on t."

Naturally, T must be the sort of Turing machine which runs on a semi-infinite tape, because in feeding the pair (T, t) to T_H, one-half of T_H's tape will hold the

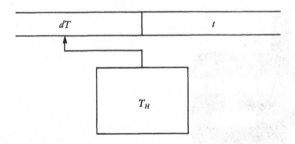

Figure 59.1   A Turing machine that solves the halting problem

description dT of T and the other half will be a duplicate of the semi-infinite tape t. A similar scheme is used to implement a universal Turing machine (see Chapter 51).

Does such a machine T_H exist?

Suppose it does. If T halts on t, then sooner or later T_H will signal the equivalent of yes and, in so doing, complete a transition from some state q_i to a halting state q_h (Figure 59.2). If T does not halt on t, however, then T_H will sooner or later say no and complete some other transition from a state q_j to a halting state q_k (Figure 59.3).

Now by carrying out a few simple alterations on T_H, we can get it into very serious trouble with itself, so to speak. The first alteration we carry out results from asking if T_H can decide whether T halts on dT, rather than t (Figure 59.4).

If T_H is asked to perform only this rather specialized (and rather strange) task, then we may supply T_H with a more compact tape containing just one copy of dT, by implanting a special Turing machine T_C within T_H. It is the business of T_C to make a copy of dT and, when it has finished, to hand matters over to T_H via a transition from T_C's final state to T_H's initial state (Figure 59.5).

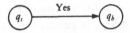

Figure 59.2   Transition to the "yes" state

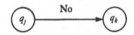

Figure 59.3   Transition to the "no" state

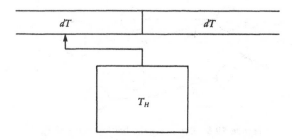

Figure 59.4 Will T halt on its own description?

Denoting the resulting machine by T'_H, we next perform a small piece of surgery on T'_H's two halting transitions. The yes transition from q_t and the no transition from q_f are both diverted into new states q_n and q_y, respectively (Figure 59.6). Once in state q_n, there is a transition back into q_n for every possible state/input combination in which T'_H might find itself. Thus, once in state q_n, the resulting maching T''_H will never halt. Once in the state q_y, however, T''_H will halt by definition: q_y is a halting state (Figure 59.7).

We may now put T''_H in a very pretty pickle by giving it the tape dT''_H to work on. If T''_H halts on dT''_H, then it must take the same yes transition which T_H would take. But in doing so, it enters a state through which it must endlessly cycle for all eternity, never halting: If T''_H halts on dT''_H, then T''_H does not halt on dT''_H! The

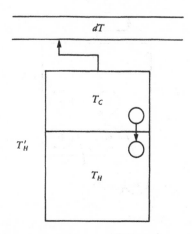

Figure 59.5 A copying machine is grafted onto T_H.

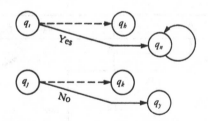

Figure 59.6 Bypassing the halting states of T_H

situation is no better if T_H'' is assumed not to halt on dT_H''. For in this case dT_H'' must take the no transition, ending in state q_y, a halting state: If T_H'' does not halt on dT_H'', then T_H'' *does* halt on dT_H''!

These contradictions ensure that machine T_H cannot have existed in the first place. Hence the halting problem is not solvable by any Turing machine.

A similar unsolvable problem is the following: Is there a Turing machine which, given any pair (T, t) as input, can decide whether T ever prints the symbol x when processing tape t? Certain alterations on a machine T_P which is alleged to solve the "printing problem" result in the same sort of contradictions as encountered above.

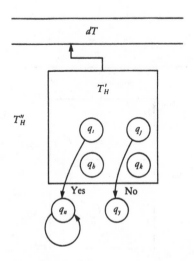

Figure 59.7 The Turing machine T_H''

Problems

1. Use the unsolvability of the halting problem to prove the unsolvability of the printing problem. Do this by altering T_P so that it solves the halting problem.

2. Besides uncomputable decision functions, there are uncomputable numerical functions. Let $\phi(n)$ be defined as 1 plus the largest finite number of 1s that any n-state Turing machine can produce from the integer n as input. Assuming that all Turing machines considered have binary alphabets, show that $\phi(n)$ is uncomputable.

References

D. Hofstadter. *Gödel, Escher, Bach: An Eternal Braid.* Basic Books, New York, 1979.

M. Minsky. *Computation: Finite and Infinite Machines.* Prentice-Hall, Englewood Cliffs, N.J., 1967.

60

COMPUTER VIRUSES

A Software Invasion

Since the early 1980s, the computing world has witnessed an astonishing onslaught of hostile programs. They have invaded thousands of microcomputers, using the disk operating system to get themselves copied and then, all too often, printing annoying messages on the screen, destroying files, and wreaking whatever havoc their perpetrators can imagine. Such hostile programs are called *viruses* because they invade from without and use the computer's "cellular machinery" (the operating system) to get themselves executed.

Figure 60.1 shows the main memory of a microcomputer as a linear column of consecutive locations. When the user of a microcomputer turns on the machine, the disk operating system or DOS (see Chapter 53) is loaded into the computer's main memory from a disk. Other files, called file A and file B in the figure, may also occupy the same disk.

When the user elects to run a particular program, say file A, the DOS searches for it on the disk and, finding it, loads it into main memory as well. If the program happens to contain a virus, as soon as it begins to execute the virus becomes "active." It may, for example, use the DOS to read in other files from

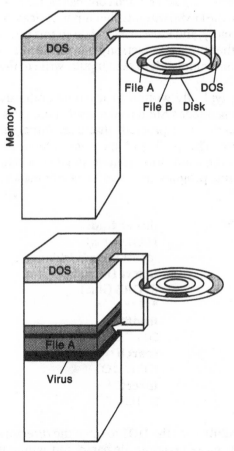

Figure 60.1 A virus infection sequence

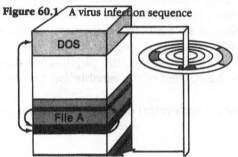

the disk and implant itself in them, as shown in the third frame of the figure. The manner in which the virus carries out these operations illustrates not only the great flexibility of "open" operating systems, but their vulnerability as well.

Disk operating systems, as well as other kinds of operating systems, include special commands for programs that use machine resources outside the context of ordinary, high-level languages. Such commands, for example, can read characters directly from the keyboard and other input devices, search main memory, search the disk memory, and move data and programs from one memory area to another, among other things. In fact, any program that runs on that operating system uses the same commands to read inputs, write to the screen or disk, and so on.

Virus programs, typically written in machine code, usually employ DOS commands to commandeer system resources that the virus must use. Here, for example, is a generic virus program that uses three subroutines, **findfile,** **search,** and **insert.** The bodies of the subroutines, indicated by rows of squares, occur after the main virus program and do not concern us. Besides the subroutines, the virus program uses two DOS commands, here called LOAD and STORE.

this := **findfile**
LOAD (this)
loc := **search** (this)
insert (loc)
STORE (this)
******
findfile
☐ ☐ ☐ ☐ ☐
search
☐ ☐ ☐ ☐ ☐
insert
☐ ☐ ☐ ☐ ☐

The subroutine **findfile** uses the DOS to open the directory of executable files (programs) on disk, picks a random file name, and assigns that name to its own variable, *this*. The virus program then uses the DOS LOAD command to bring the program thus selected, say file B, into main memory. The virus next deploys the subroutine **search** to scan the B program to find a suitable insertion site for the virus. When such a site is found, the physical memory location of that instruction becomes the content of the variable *loc*. The virus is now ready to replicate itself.

The **insert** subroutine is cleverest of all. It uses DOS to copy itself onto the end of program B, copies the instruction at *loc* to the end of the virus code, then replaces the instruction by a transfer to the virus code itself. Of course, the virus must remain invisible to ensure that the infected program still runs properly. The virus also inserts a transfer at the very end of its code back to the program B instruction immediately following the one replaced. The last virus instruction shown in the listing above uses the DOS STORE command to put the program back on the disk under the same file name. In the process, the old (uninfected) copy is effectively erased.

Figure 60.2 shows the transfers created by the virus as arrows. Readers who follow the arrows will notice that every instruction of the host program B will execute in exactly the same order as before. It will run normally. But somewhere in the middle of B's execution, the virus will become active, infecting yet another program, before returning control to B.

There are many possible points of infection for a virus that somehow gets into a DOS-based system. Recall, for example, the six stages of the booting process when such a computer is turned on (see Chapter 53).

The primary routines, stored in the read-only memory (ROM) portion of the computer hardware, cannot be infected once they are installed. After all, such memory cannot be written on. The partition record, which contains among other things the location of the disk's boot sector, can be infected by a virus, however, If there is no room for the virus in this already crowded portion of the disk, the virus may relocate the partition record to another portion of the disk,

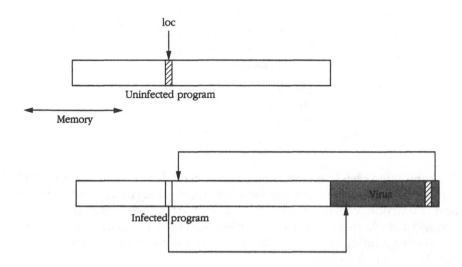

Figure 60.2 A virus permits a program to run normally

then install itself in the partition record. Every time the computer is booted, the virus runs, then transfers execution to the boot sector.

A virus may also install itself in the crowded boot sector by using essentially the same technique. It relocates the boot sector and installs itself in that site. Viruses may attack the major systems programs (such as MS.SYS or IO.SYS in the PC family of microcomputers) that come up next during the booting process. Viruses may also infect command files, getting themselves automatically executed every time the command system runs. They may also infect executable files and last, but not least, any and all user programs.

Viruses that infect the operating system at some level are far more dangerous and hard to get rid of than viruses that attack user or application programs. The latter affect only the command system, for example, when ordering it to carry out the virus's dirty work.

The infections continue from program to program to program. A virus that inhabits a disk that the user buys or borrows from someone may, in time, infect most of the programs that he or she uses. This is especially true of memory-resident viruses, infective programs that get the DOS to recognize them as a permanent parts of the operating system, in effect.

What is the point of all this infection? At some stage, a proper virus must do something beside infecting other programs. Otherwise, one assumes, there would be no emotional payoff for the software vandal who wrote the virus program in the first place. A crucial part of the virus, indicated by the row of asterisks in the illustration above, may now be shown.

day/date := **check** (clock)
if day = 5 **and** date = 13
   then bomb

check
□ □ □ □
bomb
□ □ □ □

The subroutine **check** uses DOS commands to read the system clock, and this part of the virus program assigns the appropriate values to *day* and *date*. The next instruction, called a *trigger* by virus experts, tests whether the day is 5 (Friday) and whether the date is 13. If it happens to be Friday the 13th, any copy of this virus running that day will notice that the triggering condition has been met. It will set off its **bomb**.

The **bomb** subroutine of this particular virus happens to erase all the files both on the disk and in main memory, or as far as it can get before it erases a part of the DOS (or itself) that it needs to execute properly. In any case, the user is bound to become very upset when the computer immediately ceases to function and the subsequent painful recovery reveals that some files, possibly irreplaceable, have been lost.

Viruses come in many styles and with many different missions. The styles range from *intrusive* viruses (like the one above) to *overwriting* viruses that simply obliterate a portion of the host program. The authors of the latter type may feel their mind-children to be so potent that they need to execute only once, anyway.

Virus damage varies from obliterating files or, worse yet, altering them in subtle ways, to printing annoying messages on the screen. There is even one virus that does nothing at all but propagate, its author a relatively gentle soul suffering only from a fascination with reproductive power.

Triggers vary from days or dates to other conditions prevailing in a computer's operating system at any moment. It could simply be the presence of a certain file name on disk. The virus's mission: to destroy all files by that name, wherever it finds them.

Computer scientists, like Americans Fred S. Cohen and Eugene Spafford, have classified viruses and have reproduced them in the "laboratory," the better to study them. Their long-range prognosis, summarized in the "law" below, implies an endless future battle between technopaths and those who would make systems accessible yet secure:

For every system there is a way in, for every virus there is a defence.

The student of vandalware quickly learns the fundamental distinction between viruses and *worms.* Viruses depend utterly on a programming environment such as a DOS to spread. Worms are independent programs that typically infect not microcomputers but mainframes. The famous worm that spread throughout the Internet in November of 1988, for example, traveled independently from computer to computer along standard communication lines. Once set operating in a single computer of the network, the worm spread by examining the list of remote users, getting their names, dialing up their home machines and then guessing their passwords.

The worm enjoyed great success for only two reasons. The UNIX operating system that ran many of the Internet computers had many insecure features that permitted users great flexibility but also permitted the worm easy access. The second reason lay in the unimaginative passwords dreamed up by many of Internet users. When guessing passwords, the worm would simply consult its long list of words from "aardvark" to "zoom."

Problems

1. Do NOT write a virus or a worm program.

2. A program stored in executable form is infected by a virus but, when the program runs, it detects the virus and reports the infection via a message to the screen. How might the program detect the virus?

3. Design a virus-detection program module that can be installed in any executable program.

References

Lance J. Hoffman. *Rogue Programs: Viruses, Worms and Trojan Horses.* Van Nostrand Reinhold, New York, 1990.

Ralf Burger. *Computer Viruses: A Hi-Tech Disease.* Abacus Books, Grand Rapids, Mich., 1989.

61

SEARCHING STRINGS

The Boyer-Moore Algorithm

There are unexpected subtleties in the seemingly straightforward matter of searching a string of characters for a particular pattern, especially if we want to do it quickly. Such an operation is carried out routinely by text editor and word processing programs. Thus the question

> How fast can you search a string of *n* characters
> for a given pattern?

has a certain practical pungency.

For example, as every prospector knows, there are many ways of searching for GOLD in THEM THAR HILLS. A reasonable amount of method in our madness would dictate that we search the string in some definite order, say, left to right. Thus we will attempt to match the pattern with the leftmost part of the string and, if it fails to match, continually shift it to the right until either it does match or we run out of string. It seems equally reasonable to compare characters in the pattern and the string in left-to-right order:

> THEM THAR HILLS
> GOLD

Since the G and T do not match, however, there is no point in further comparison, and we shift the pattern 1 unit to the right and try again.

> THEM THAR HILLS
> GOLD

Again, the initial characters do not match. Continually shifting in this way, we eventually run out of string and conclude that there is no GOLD in THEM THAR HILLS. Some text editor programs carry out string searching in precisely this way.

An algorithm discovered by R. S. Boyer and J. S. Moore in 1977 injects a little madness into this method by matching characters from right to left instead. If the string character being matched is not in the pattern, one may shift the pattern by its entire length. Thus from

> THEM THAR HILLS
> GOLD

we go immediately to

> THEM THAR HILLS
> GOLD

At first sight, this seems a dramatic increase in efficiency, especially since in the next step the pattern can be jumped another four characters for the same reason. Unfortunately, this increase in efficiency is only illusory, since to check that A is not in GOLD requires as many comparisons as shifting GOLD ahead one space at a time after a comparison of G with each of four characters in the string.

If, however, there is a table, say *table 1*, which for each letter of the alphabet contains its rightmost position in the pattern (GOLD), then it is only necessary to look up A in the the table (this can be done very quickly) and note that its corresponding entry is $\varnothing$. One may then shift the pattern to the right by as many characters as are in it.

Besides this technique of shifting based on *table 1*, the Boyer-Moore algorithm uses one other neat idea: Suppose that in the process of matching the pattern characters with those of the string, the first m characters (from right to left) are found to match:

$$PETERPIPERPICKEDAPECKOF \overline{PICK}LEDPEPPERS$$
$$P\ I\ C\underline{NIC}$$

In such a case, the portion so far matched at the end of the pattern may well occur elsewhere in the pattern. It would then be wise to shift the pattern to the right only by the distance between the two portions

$$PETERPIPERPICKEDAPECKOF\overline{PICK}LEDPEPPERS$$
$$\rightarrow\ PIC\underline{NIC}$$

In the example above, the Boyer-Moore algorithm would now quickly discover that E is not in PICNIC and, as a result, shift the pattern by another six characters.

The Boyer-Moore algorithm is short and straightforward. Besides *table1*, it uses another table, *table2*, which for each possible terminal portion of the pattern indicates the rightmost (nonterminal) recurrence of it. This table can also be accessed quickly.

STRING

1. $i \leftarrow width$
2. **while** string remains
    1. $j \leftarrow width$
    2. **if** $j = 0$ **then** *print* "Match at" $i + 1$
    3. **if** $string(i) = pattern(j)$
        then $j \leftarrow j - 1$
        $\qquad i \leftarrow i - 1$
        go to 2.2
    4. $i \leftarrow i + max(table1(string(i)), table2(j))$

where $width$ = length of pattern
$\qquad length$ = length of string
$\qquad\qquad i$ = position of current string character
$\qquad\qquad j$ = position of current pattern character
$\qquad string(i)$ = ith character in string
$\qquad pattern(j)$ = jth character in pattern

As an example of the algorithm STRING in operation, suppose PICNIC has just been placed next to KOFPIC as in the first PETERPIPER . . . string above. Starting at statement 2.3, we see that $i = 26$ and $j = 6$. Since $string(26) = pattern(6)$, both i and j are decremented. By returning to statement 2.2 with

$i = 25$ and $j = 5$, another character match is found, and i and j now become 24 and 4, respectively. The characters in the latter positions do not match, however, and execution of the last line proceeds:

$$string(24) = P$$

and *table*1 looks like this

N	2
O	6
P	5
Q	6
R	6

Thus, $table1(P) = 5$, meaning that the i pointer must be shifted at least 5 characters to the right before new matching attempts have any hope of success. This is equivalent to shifting the pattern 3 characters to the right, where the P's will then coincide.

However, *table*2 looks like this:

1	11
2	10
3	9
4	5
5	4
6	1

Again, the i pointer must be shifted at least 5 characters to the right before it is reasonable to restart the matching process.

As it happens in this example,

$$table1(P) = table2(4)$$

and the i pointer is shifted 5 characters in any case. However, trying the patterns PXXNIC and XICNIC will convince the reader of the independent usefulness of these tables.

The Boyer-Moore algorithm has been exhaustively tested and rigorously analyzed. The results of the tests show that on the average, the Boyer-Moore algorithm is "sublinear." That is, to find the pattern in the ith position of the string requires fewer than $(i + width)$ instructions when the algorithm is suitably coded. In terms of worst-case behavior, the algorithm still executes in the order of $(i + width)$ instructions.

As far as the relative usefulness of *table*1 and *table*2 is concerned, it is interesting that with large alphabets and small patterns *table*1 is most useful. However, this situation reverses itself for small alphabets and large patterns.

Problems

1. Write the Boyer-Moore algorithm in the language of your choice, and test its sublinearity by counting the number of iterations of the **while** loop and comparing this number with the sum of string and pattern size.

2. Devise a pattern and an infinite set of strings on each of which the Boyer-Moore algorithm takes linear time (in the sense of Problem 1) to search for the pattern.

References

Donald E. Knuth. *The Art of Computer Programming*, vol. 3. Addison-Wesley, Reading, Mass., 1967.

Thomas A. Standish. *Data Structure Techniques*. Addison-Wesley, Reading, Mass., 1980.

62

PARALLEL COMPUTING

Processors with Connections

magine n simple computers (called *processors* in this chapter) arranged in a row or an array and connected in such a manner that each processor may exchange information with only its neighbors to the right and left. The processors at either end of the row are used for input and output. Such a machine constitutes the simplest example of a systolic array. It can solve many problems much more quickly than a single processor. One of these is the famous *n-body problem* of physics: Compute the path of each of n bodies moving through space under the influence of their mutual and combined gravitational attractions.

A sequential computer can carry out the computation of all $(n^2 - n)/2$ attractions in $O(n^2)$ basic steps. A systolic array (Figure 62.1), however, can achieve the same result in $O(n)$ basic steps, an n-fold improvement in speed! Here is how it works.

The basic cycle of an n-body computation involves calculating, for each of the bodies, the summed attractions of the other $n - 1$ bodies. It is reasonable to assume, for the systolic array above, that each processor carries a program to

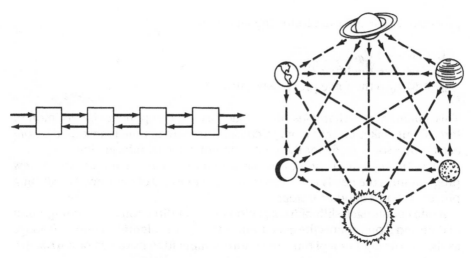

Figure 62.1 A systolic array tackles the planets

calculate the newtonian force formula:

$$F_{ij} = km_i m_j / d_{ij}^2$$

where m_i and m_j are the masses of the ith and jth bodies, respectively, k is the gravitational constant, and d_{ij} is the distance between them. At each stage of the computation, each processor, given the coordinates of two bodies, computes the distance between them by the standard euclidean formula and then computes the force as given above.

The computation begins when the coordinates and mass of the nth body B_n are fed to the first processor P_1. The coordinates and mass of B_n are passed along to the second processor P_2, just as the coordinates and mass of the $(n-1)$st body B_{n-1} are inputted to P_1. This process continues until each processor holds the coordinates and mass of a unique body in the gravitational ensemble. In general, P_i holds the coordinates and mass of B_i. Obviously, this process absorbs n transmission cycles.

The second phase of the systolic array computation involves exactly the same input sequence. But now each processor holds the coordinates and mass of its assigned body. As the information for each new body B_j arrives from the left, the processor P_i executes the following algorithm:

1. Calculate d_{ij}.
2. Calculate the force F_{ij}.
3. Add F_{ij} to the force previously calculated.

This summary oversimplifies much (such as the componentwise addition of forces) but makes clear that only a constant amount of time is absorbed before each processor is ready for the next round of celestial information.

After $2n$ steps, each processor has accumulated $n - 1$ forces, and a new program shifts the newly computed forces out one end of the array. The shifting process absorbs another n steps.

If one counts each shift of information in either direction as absorbing 1 unit of time and if one counts the execution of the simple algorithm outlined above as also absorbing 1 unit of time, the entire computation has taken only $4n$ steps, a great improvement over the sequential (nonparallel) computer.

Systolic arrays may be much more sophisticated than the simple model used for the n-body calculation just described. For example, a commonly discussed geometry of processors is a square or rectangular grid. In the context of parallel computers, likewise, systolic arrays are just one of a vast range of quite distinct types either already constructed or still in the planning stages. Representative of the more general and powerful schemes is the cube-connected computer.

A d-dimensional hypercube (see Chapter 49) forms the basis for connections between n processors in the cube-connected computer. Here, each processor occupies a vertex of the cube. The total number n of processors is therefore 2^d. Of course, this arrangement refers only to the topology of connections between the processors, not to the actual geometry. To emphasize this point (which applies equally well to systolic arrays), one may arrange all n processors in a square (Figure 62.2).

We will use a cube-connected computer on an important practical problem discussed earlier for sequential machines, namely matrix multiplication (see Chapter 25). Given two $n \times n$ arrays X and Y, how quickly can we form the n^2 elements in the product array Z? In Chapter 25, the state of sequential art was left hovering somewhere between $O(n^2)$ and $O(n^3)$. Here the time will be reduced to just $O(\log n)$ steps!

Up to a certain point, it helps to throw additional processors at certain problems. To carry out matrix multiplication at such blinding speed, n^3 processors will be required. Here again, it does no harm to assume that n^3 is itself a power of 2. In any event, before we begin an examination of the parallel matrix multiplication algorithm, it is worthwhile to observe that up to d separate links might be required so that one processor might communicate with another. In the case of n^3 processors inhabiting the d-cube, $d = \log n^3 = 3 \log n$. In other

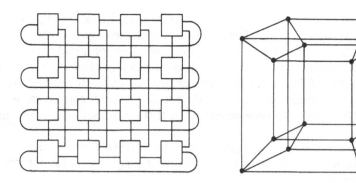

Figure 62.2 A hypercube in two forms

words, mere communication between processors will require $O(c \cdot \log n)$ time steps, where c is a "communication constant," hereafter ignored.

The i,jth element of the product matrix Z has a simple formula:

$$z_{ij} = \sum_{k=1}^{n} x_{ik} y_{kj}$$

By coincidence, the parallel computation proceeds in three phases here just as it did in the systolic array example. The first phase distributes the array elements from X and Y throughout the n^3 processors, the second phase computes the products, and the third phase forms the sums.

It is convenient to index each processor by a triple of numbers (k,i,j). Each index runs through n consecutive binary values. Each processor $P(k,i,j)$ will be connected to every processor that differs from it by 1 bit in exactly one of its index values. Thus, processor $P(101,i,j)$ is connected to processor $P(001,i,j)$.

Initially, the entries x_{ij} and y_{ij} are deposited in processor $P(0,i,j)$. The first phase then distributes these entries throughout the cube-connected computer so that, in general, $P(k,i,j)$ contains x_{ik} and y_{kj}. The method is described here for the case of x_{ik} only, but the communication of y_{kj} is similar.

First, the x_{ik} content of $P(0,i,k)$ is transmitted to $P(k,i,k)$ by a route that takes it through fewer than $\log n$ other processors. The message itself consists of the number x_{ik} and the target processor address (k,i,k). At each stage of the message's journey, the next processor it goes to is the one having an index that is 1 bit closer to k. For example, if $k = 5 = 101$, then the sequence followed might be

$$P(000,i,k) \rightarrow P(100,i,k) \rightarrow P(101,i,k)$$

Next, when each processor $P(0,i,k)$ has sent its message to the target processor $P(k,i,k)$ (all in parallel), each of these sends the same message to all processors $P(k,i,1)$, $P(k,i,2)$, . . . , $P(k,i,n)$. This is done by a form of broadcasting. The message is sent simultaneously through all communication lines in which one of the relevant bits differs from the current bit. For example, if $k = 101$, as above, the message might be sent according to the parallel pattern shown in Figure 62.3. Here again, the time absorbed by this transmission is fewer than $O(\log n)$ steps.

With the matrix entries x_{ik} and y_{kj} in processors $P(k,i,j)$, the second phase of the parallel computation takes place: The product $x_{ik} \cdot y_{kj}$ is formed and stored in the same processor.

The third phase, like the first, is slightly complicated. Essentially, the products are drawn from all the processors $P(1,i,j)$, $P(2,i,j)$, . . . , $P(n,i,j)$ into the single processor $P(0,i,j)$ by a continual "fanning in" of accumulated sums. Again, a message is transmitted from processor to processor, but always in a direction that changes a 1 bit in the first index to 0. When two such sums arrive at the same processor, they are added to the product already in the processor and then retransmitted. Suppose, for example, that $n = 4$ and that the products 8, 7, 5, 3, 9, 12, and 6 have just been computed in $P(1,i,j)$ through $P(7,i,j)$, respectively. Figure 62.4 shows a set of possible routings for the sum information to reach $P(0,i,j)$. In this phase of the computation, all transmission and all summing are carried out in parallel, with the number of steps bounded merely by the maximum distance between two processors, namely, $\log n$.

In this manner the product of two $n \times n$ matrices is completed in just $O(\log n)$ steps.

The prospect of parallel computing has spurred new developments in the field of algorithmic analysis. Parallel algorithms now exist for almost every problem classically studied. A reasonable framework in which to study parallel algorithms involves the determination for each problem of whether it belongs

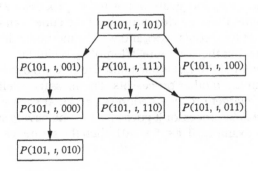

Figure 62.3 Distributing matrix elements

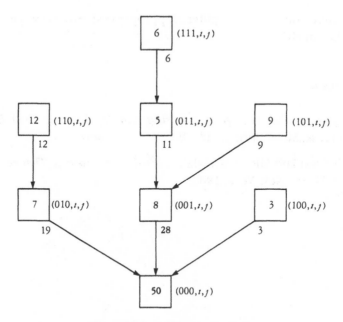

Figure 62.4 Adding up the results

to a set called *Nick's class* (NC). (The "Nick" comes from Nicholas Pippenger, a computer scientist at IBM's San Jose Laboratory.) To qualify for membership, a problem must be solvable in polylog (the polynomial of a logarithm) time by a polynomial number of processors.

Specifically, this implies the following ground rules: Given a certain problem P, there must be polynomials p and q such that for an instance x of P having size n, there is an algorithm that, running on $p(n)$ processors, will solve x in time $q(\log n)$. We have just seen an example of such a problem — matrix multiplication carried out on n^3 processors required on the order of $\log n$ steps to solve. In this case p is a cubic polynomial, and q is a linear one.

Problems

1. Let P be a problem for which the fastest possible sequential algorithm requires $O(f(n))$ steps. Show that a parallel computer with a *fixed* number of processors cannot solve the problem more quickly.

2. Use a linear systolic array to sort n numbers in time $O(n)$.

3. Use a cube-connected computer of appropriate size to show that the sorting problem lies in NC.

References

Kai Hwang (ed.). *Supercomputers: Design and Applications.* IEEE Computer Society Press, Silver Spring, Md., 1984.

G. Jack Lipovski and Miroslaw Malek. *Parallel Computing: Theory and Comparisons.* Wiley, New York, 1987.

THE WORD PROBLEM

Dictionaries as Programs

The idea of looking up a word in a dictionary is simple enough, but it leads directly to a problem that cannot be solved by a computer. The problem is based on the equivalence of strings. For example, given a sentence in English, choose a word at random, use a dictionary to select an equivalent word, and then substitute the new word for the old one. When are two sentences equivalent under a sequence of such operations?

The question seems trivial when we imagine two sentences in English. Of course we know when two sentences are equivalent: They only have to have the same meaning. But a computer has no such advantage. So far no one has constructed a program that understands meanings.

It is easier to appreciate the problem computationally if we change the language, say, to ancient Martian. Figure 63.1 shows a fragment of parchment retrieved from a cave at the foot of Olympus Mons. Another fragment gives a portion of the Martian dictionary. Enough of the dictionary remains to determine whether the sentences are equivalent. Martian sentences have no breaks in them; there is no space character. Are these two sentences equivalent? To find out, systematically make substitutions for words in the top sentence in the

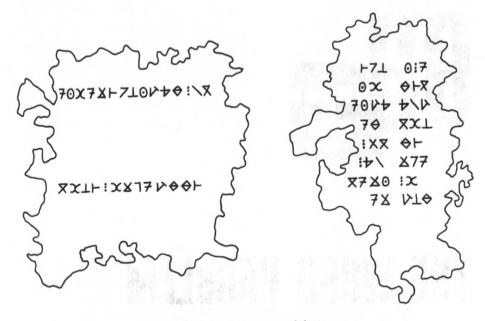

Figure 63.1 A Martian text and dictionary

hope of ultimately obtaining the bottom one. Of course, only those substitutions shown in the dictionary fragment are available. If a word in the left-hand column is located in the current sentence, substitute the corresponding word in the right-hand column for it. At first sight, this procedure seems sound and straightforward, but it has a flaw. In fact, there is no procedure that works.

In 1914 Axel Thue, a Norwegian mathematician, made the first formal statement of this problem: Let Σ be a finite alphabet of arbitrary symbols, and let D be a dictionary consisting of a finite number of pairs (X_i, Y_i) of words. Given an arbitrary word X over the alphabet Σ, a substitution involves finding a subword X_i in X and substituting the corresponding word Y_i. The words X and Y are equivalent if there is a finite sequence of substitutions taking X into Y.

A humbler and somewhat more readable example of this problem uses the Roman alphabet:

RTCXUPNTRX	PN	XXR
	NTR	CU
	XUPN	CXXP
	TCX	RNC
	RX	NP
	PRC	UTC
	TU	XRN
UUTCXXRNP	NTT	CC

416

Is it possible to convert the word at the upper left to the word on the lower left by a sequence of dictionary substitutions? The exercise might tire a human reader, but surely a computer could be programmed to determine the equivalence of these words. The method implicit in the brief algorithmic description above would start with the upper word and derive all possible substitutions in it (Figure 63.2).

Maintaining a list of all words currently derived, the program simply iterates this same procedure for every word on the list. Words in which substitutions are made are replaced on the list by the resulting words. As each new word is added to the list, the computer tries to match it with UUTCXXRNP. If the latter word is equivalent to the former, this program will certainly find a match, sooner or later.

The last sentence gives the impression that the problem has been solved. But it has not. Even if the two words were equivalent, the sequence of substitutions taking one word into the other might involve intermediate words that are arbitrarily long. There is, consequently, no way of knowing when the computation should terminate. If the two words are not equivalent, the computer could run forever as we patiently await the results.

One of the cardinal features of computability is that a computation must sooner or later halt in all cases. To be computable, a function (even that of yes or no) must be computable in finite time.

The failure of the search algorithm just described is no proof that the word problem (also called the *word problem for semigroups*) is unsolvable by computer. After all, how do we know there is not some subtle theory which can be implemented in a far different algorithm? Perhaps such an algorithm might even run in linear time; it might determine the equivalence of two strings of length at most n in $O(n)$ steps!

That no algorithm of any description exists for this problem is best seen by invoking a very old mathematical trick: Develop a transformation between the problem and one about which something is already known. The transformation is called a *Turing reduction* and is closely akin to the polynomial-time transformations described in Chapter 54.

Since the difficulty encountered by the naive algorithm was in halting, perhaps Thue's word problem is related to the halting problem for Turing ma-

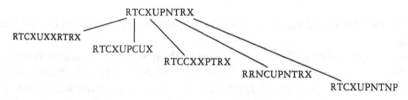

Figure 63.2 Beginning the word-generation process

chines (see Chapter 59). This guess turns out to be correct. To see how, we must be able to convert a Turing machine to a Thue problem. First, we make an explicit statement of both problems. Respectively, they require one to find algorithms that do the following: Given an arbitrary Turing machine and initial tape, the first algorithm determines whether the machine will halt. Given a dictionary and a pair of words, the second algorithm determines whether the words are equivalent. The parallelism of these two statements leads us to wonder whether the reduction will involve replacing a Turing machine by some kind of dictionary. The key notion leading to just such a replacement is the so-called instantaneous description of a Turing machine and its tape environment. Merely copy the entire nonblank portion of the Turing machine's tape. Then, to the left of the cell currently being scanned, insert a new cell containing a symbolic name for the machine's current state. For example, if the Turing machine's current tape looks like

0	1	1	#	0	#	0	0	0	1	#	0	1	1	0

suppose the machine is scanning the 10th square from the left and is in state 5. Replace the symbol in that square by some symbol not in the tape alphabet, say F (for "five"). Now the tape looks like

0	1	1	#	0	#	0	0	0	F	1	#	0	1	1	0

It should now be clear how to proceed. Construct a dictionary that reproduces the local effect on such a tape of the Turing machine's program when applied to it at the current read/write head position.

If, for example, when the machine is in state 5 and reads a 1 it enters state 2, writes a 0, and moves one square to the left, the corresponding dictionary entries would be the following:

0F1	T00
1F1	F10
#F1	S#0

In this case, the first of these entries would apply. Replace the cell contents 0F1 by T00.

The process just described, however, is nothing more than an example of the substitutions used in Thue's word problem! The complete reduction, unfortunately, involves a little more apparatus. For one thing, two words must be supplied if we are finally to arrive at a bona fide instance of the word problem.

Let the first word be the initial tape with the state marker already introduced to the left of the starting cell.

It may seem overly restrictive, but it is perfectly all right to allow the second word to be entirely 0s, or even just a single special symbol, say Z. Any Turing machine can be modified to produce this effect if it halts at all.

A moment's thought makes it clear that the reduction is Turing-computable. For it to be Turing-computable means simply to be computable (see Chapter 66). The two words and the dictionary can easily be computed by an algorithm which takes a Turing machine's program and initial tape as input.

There is no algorithm to solve the halting problem. We have, however, a computable transformation of that problem to the word problem. If the latter were solvable, the former would be, too.

Problems

1. Given an arbitrary Turing machine, modify it so that it uses special marker symbols to delimit visited cells of its tape from those not yet visited. If it has a halt state, show how the addition of a few more states and lines of program suffices to have it convert its entire "final" tape (including the markers) back to blank cells. When a new halting state is finally entered, it prints a Z.

2. A word ladder is a recreation in which two English words of the same length are given to a player. Can one word be converted to another by single-letter substitutions? The "dictionary" in this case consists of all possible letter pairs. The game would be trivial were it not for the fact that all intermediate strings must be English words. Is there an algorithm for generalized word-ladder play? The input is an alphabet Σ, a language L over Σ, and two words in L. All single-symbol substitutions are allowed, but all intermediate words must lie in L.

3. In specific instances of Thue's word problem, it may be possible to show that no solution exists. Do this for the second example in this chapter.

References

S. C. Kleene. *Introduction to Metamathematics.* Van Nostrand, Princeton, N. J., 1950.

Harry R. Lewis and Christos H. Papadimitriou. *Elements of the Theory of Computation.* Prentice-Hall, Englewood Cliffs, N.J., 1981.

64

?

LOGIC PROGRAMMING

Prologue to an Expert System

Traditional programming languages enable the programmer to solve problems by specifying computational routes to solutions. A new generation of languages makes much computation implicit merely by specifying certain properties that a solution must have. Traditionally, we have sent young children to buy ice cream by giving them directions to the store. Now we need only describe the ice cream, so to speak.

The new generation of languages involves logic programming. First-order predicate calculus (see Chapter 58) forms the basis of their operation. Best known of the logic programming languages is Prolog, short for *pro*gramming in *log*ic. Invented by a group of scientists in Marseilles, France in 1976, Prolog has been proposed as the fundamental language for a whole new breed of parallel computers referred to as the *fifth generation*.

The logical powers of Prolog can be illustrated in a simple way by probing a family tree (Figure 64.1). A Prolog program consists of clauses just like expressions in the predicate calculus. Some clauses are instances (i.e., contain constants but no variables) while others contain variables and are quantified. Con-

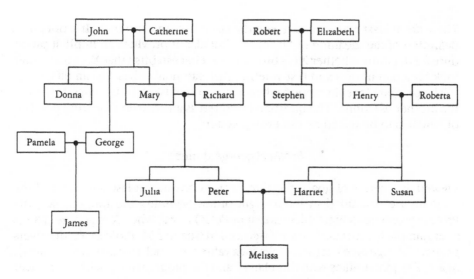

Figure 64.1 A genealogical chart

sider for example, the statement

female(mary)

This is an instance or unit clause. However, the statement

brother(*X, Y*):-*male*(*X*), *parent*(*Z, X*), *parent*(*Z, Y*)

represents a quantified clause. It could be read as follows: For all *X, Y,* and *Z*, if *X* is a male, *Z* is a parent of *X*, and *Z* is a parent of *Y*, then *X* is a brother of *Y*. Written in logical notation, it has the form

$$\forall X, Y, Z \; (male(X) \wedge parent(Z,X) \wedge parent(Z,Y)) \rightarrow brother(X,Y)$$

We are already in a position to write a simple Prolog program. First, we list a number of instances, each consisting of a single Prolog statement:

parent(john,donna).
parent(john,george).
parent(john,mary).
female(donna).
male(george).
female(mary).
brother(*X,Y*):-*male*(*X*), *parent*(*Z,X*), *parent*(*Z,Y*).

There are at least two ways to regard the last line of this program. It is not only a definition of the meaning of "brother," but also it provides an implicit procedure for deciding whether X is a brother of Y. First establish that X is a male, then look for a parent of X and test whether that parent also has Y as an offspring.

The program is set in motion when a question is asked by the user. Is George the brother of Donna? The question is written as an assertion the truth or falsity of which is to be tested by the Prolog system:

$$brother(\text{george,donna})$$

Viewed as a series of goals, the three conditions in the last statement of the program are examined in left-to-right order. Scanning the unit clauses, the Prolog system quickly establishes that $male(X)$ is true when $X =$ george. This is a very simple form of unification discussed in Chapter 54. Prolog next discovers $parent(\text{john,george})$. It then looks for a value for Z such that $parent(Z,\text{donna})$. Again, $Z =$ john unifies with this clause, and the program quits with the answer yes—all three goals have been met.

The powers of Prolog go far beyond this sample computation. More than yes or no answers are supplied for one thing, and more complicated forms of program are possible. Suppose, for example, that we want to find out the names of James's cousins. The first part of the Prolog program is assumed to contain all the relevant unit clauses such as $parent(\text{mary, julia})$. There may be unit clauses which are no help at all in providing an answer. In Prolog programs, one cannot always be sure which unit clauses will be useful for a given computation.

The main part of the program consists of the quantified clauses:

$$sibling(X,Y):- ne(X,Y), parent(Z,X), parent(Z,Y).$$
$$cousin(X,Y):-parent(W,X), parent(Z,Y), sibling(W,Z)$$

Finally comes the question:

$$cousin(\text{james},C).$$

Prolog attacks the problem by searching the clauses for one containing the *cousin* predicate. Unifying through the substitution $X =$ james, *Prolog* produces a new, intermediate clause which it stores in memory:

$$cousin(\text{james},C):-parent(W,\text{james}), parent(Z,C), sibling(W,Z).$$

As before a series of goals is set up:

1. Find a parent W of James.
2. Find a parent Z of C.
3. Find a sibling Z of W.

The moment all three goals are satisfied for a given set of values for W, C, and Z, Prolog will output the value it has found for C since that appeared, implicitly, as a question to be answered.

Prolog uses backtracking, where necessary, to answer such questions. At first, it scans the list of unit clauses, seeking a parent of James. It finds $W =$ pamela and then operates on the new goals:

2. Find a parent Z of C.
3'. Find a sibling Z of Pamela.

It now scans all the parent clauses, finding many (Z,C) combinations that achieve goal 2 in the process. But for each value of Z thus found, Pamela does not turn out to be Z's sibling; there is simply no indication in the unit clauses of who Pamela's siblings are. Prolog then backtracks to the next value W it finds that satisfies

$$parent(W, \text{james}).$$

This turns out to be george. Thus $W =$ george, and 3' is replaced by a newer goal:

3''. Find a sibling Z of George.

Prolog again scans the parent clauses and soon discovers a parent Z of C who also satisfies $sibling(Z,\text{george})$, namely, $Z =$ john. In this case, unfortunately, C happens to be Donna and since Donna is the parent of no one, the computation thus fails. There is Mary yet to try for the value of C, however, and when Prolog encounters the unit clause $parent(\text{mary, julia})$, it then has values for W, C, and Z that satisfy all the goals implied by the last predicate. It prints the value of C as "Julia." If all cousins are wanted, the program is restarted, and it almost immediately finds the other cousin, Peter, of George.

The three predicates comprising the last clause were not the only ones to initiate the goal-forming process. Throughout the foregoing computation, Prolog continually tested

$$ne(X,Y), parent(Z,X), parent(Z,Y)$$

in left-to-right order. The first of these predicates belongs to the Prolog system: It is true only when X and Y do not have the same value. Thus, by the time Prolog is testing Z for parenthood of X and Y, the latter variables never have the same value.

Note that the order in which one writes Prolog clauses can affect the speed of the computation. It may be better, in fact, to place the clause defining *cousin* ahead of the one defining *sibling*. The latter clause has a greater number of satisfying instantiations than the former, and most of these are dead ends. It is better to encounter failed computations earlier, rather than later, in the course of Prolog's search for a solution.

With this modest understanding of the spirit of Prolog, we are ready to illustrate its role in expert systems.

Broadly defined, an *expert system* applies a certain degree of reasoning ability (typically by the use of logic programming) to a set of rules and a data base that comprise what might be called *expert knowledge.* The user of such a system may ask it questions within the domain of the program's expertise and expect useful answers.

Expert systems have been constructed for an amazing variety of human knowledge domains. There are expert systems which do a reasonable job of diagnosing infections, evaluating sites for mineral exploration, repairing machinery, and deciding legal questions based on statutes. In each case, the knowledge base of particular facts useful to the system is embedded in a large number of unit clauses. The kinds of deductions that an expert might make based on available evidence are embedded in predicate clauses of the kind used to define *sibling* and *cousin* in the previous example. Questions can then be framed within the domain of discourse that is defined by such clauses.

Even in the humble domain of human blood relationships, it is possible to imagine an expert system in operation. Suppose, for example, that a marriage is proposed under the canon law of the Roman Catholic Church. Such a marriage is forbidden, or considered void if enacted, when the parties have less than the fourth degree of consanguinity by canonical computation. Such a computation involves the identification of a common ancestor who is three generations (or less) removed from both descendants. In short, the marital hopefuls are barred from marriage if they have either a common parent or a common grandparent. Since the former case implies the latter, it is enough for an expert system on canonical marriage law to be supplied with the following clauses:

$grandparent(X,Y) :- parent(X,Z), parent(Z,Y)$
$void(X,Y) :- grandparent(Z,X), grandparent(Z,Y)$

To these clauses would be added a host of others including cases involving previous marriages still in effect, a female less than 12 years old or a male less than 14, a male who is physically impotent, and either a male or a female who has taken a previous vow of chastity. The list is much longer than this. Included in such an expert system might be those inpediments to marriage which may be removed by a papal dispensation.

In truth, such an expert system, when complete, would not require great reasoning powers. The great majority of predicate clauses would have a form no more complicated than

$void(X,Y):-chaste(X)$

or, at worst,

$void(X,Y):-ne(Y,Z),\ married(X,Z).$

Indeed, such simplicity is very common in current expert systems. But the term *expert system* is sometimes used loosely enough to include chess-playing programs. The expertise of such programs is rarely embedded in a logic programming system mainly because on sequential machines (still currently in the majority) logic programs tend to run rather slowly. Their natural milieu is the parallel computer in which a multitude of branches in the tree of possibilities can be simultaneously tested. (See Chapter 62).

Problems

1. Use Prolog syntax to define predicates *niece, great-uncle,* and *second cousin.*

2. Assuming the ready availability of unit clauses of the form *divides*(X,Y), *greater*(X,Y), and *not equal*(X,Y), write a short Prolog program that computes the greatest common divisor of two numbers.

3. Devise an expert system (in the form of a Prolog program) that plays unbeatable tick-tack-toe. Label the nine positions on the playing surface, and set up predicates such as $X(4)$, meaning the X side has played its token in position 4. Assume that new moves are added by the system to the "data base" of unit clauses and that the system is to respond with a move that avoids defeat. What predicate clauses will make such a system work?

References

Avron Barr and Edward A. Feigenbaum, eds. *The Handbook of Artificial Intelligence,* vol. II. Heuris Tech Press, Stanford, Calif., 1982.

Larry Wos et al. *Automated Reasoning: Introduction and Applications.* Prentice-Hall, Englewood Cliffs, N.J., 1984.

65

RELATIONAL DATA BASES

Do-It-Yourself Queries

I n many applications involving the storage of information in computer files, there is a single query that the user of the file wishes to make. For example, the file might be a telephone directory. To find someone's telephone number, the user of such a system merely types the person's name on the keyboard, and moments later the corresponding telephone number appears on the screen. The computer, using the name as a key, has searched through the file for that name, retrieved the telephone number stored with the name, and then displayed the number on the screen.

But some forms of data are more complex. They involve many kinds of information with special relationships between them. Consider, for example, the data associated with a relatively simple operation like a soccer league. There are relationships between players and teams, between players and telephone numbers, between teams and coaches, and so on (Figure 65.1). By using a relational data base, it is possible to produce information that is not stored in immediately accessible form. There might be a list of players, their teams, and telephone numbers, but no list of telephone numbers for all the players of the

Figure 65.1 Data relations in a soccer league

Moose Jaw Breakers. (If their game has been canceled, it would certainly be useful to have such a list.)

In a relational data base, it is possible to frame a query to produce such a list—and much more. To be specific about the range of possible queries in a relational data base, suppose that information about the soccer league is stored in three separate tables:

Table of Coaches

Team	Coach	Telephone
Breakers	Henry Hazer	731-5856
Loons	Don Sackcloth	356-2782
⋮	⋮	⋮

Table of Players

Player	Age	Telephone	Team
Brian Footloose	23	488-4782	Breakers
Bruce D. Head	31	374-4677	Chainsaws
⋮	⋮	⋮	⋮

Table of Games (the Schedule)

Home	Visitor	Date	Time	Place
Chainsaws	Blizzard	Oct. 5	2:00	Hemlo Field
Loons	Breakers	Sept. 22	2:00	Granite Park
⋮	⋮	⋮	⋮	⋮

Each table represents a relation that will be used by the data base. Each relation, moreover, is actually a set of tuples; for example, the first relation consists of triples having the form

Coaches: (team, coach, telephone)

The other two relations consist of 4- and 5-tuples, respectively:

Players: (player, age, telephone, team)
Games: (home, visitor, date, time, place)

Each column in a table (relation) represents an attribute. Some attributes act as keys when tables are searched. In the three tables just displayed, some key attributes might be *team, player,* and *home, visitor*.

Mathematically speaking, a *relation* is a set of n-tuples $(x_1, x_2, \ldots, x_n)$ where each element x_i is drawn from a set X_i (and no tuple is repeated). The tables in a relational data base satisfy this definition. As a result, mathematical theory applies; a theory called *relational algebra* defines operations on the tables.

Two such operations are *selection* and *projection*. The selection operation specifies a subset of rows in a table by means of a boolean expression involving attributes. Projection specifies a subset of columns in a table by listing the attributes involved. The projection operation also eliminates duplicates from the resulting rows.

To produce a telephone list for the Breakers, these two operations suffice. Selection is denoted by σ and projection by π:

$$\pi_{\text{player and telephone}} \left(\sigma_{\text{team=Breakers}}(\text{Players}) \right)$$

Working from the inside out, the selection operator σ specifies all rows of the players table in which the team name is Breakers. Next, among these, the projection operator π specifies the attributes *player* and *telephone,* discarding

429

all other attributes of the 4-tuples specified by π. The following list would thus appear:

Brian Footloose   488-4782
Hans Orff   482-6363
⋮   ⋮

Another important relational operation is called the *natural join*. Denoted by the symbol $\bowtie$, it operates on two tables having one or more common attributes. For each pair of tuples (one from the first table, one from the second), it produces a new tuple if the common attributes have identical values for the pair. Suppose that the relational data base system is asked to form the join of Coaches and Players:

Coaches $\bowtie$ Players

The two tables have *team* as a common attribute. One entry in the new table (i.e., relation) resulting from the join operation would be

Breakers, Henry Hazer, 731-5856, Brian Footloose, 23, 488-4782

The attribute *team,* having the common value Breakers for one row from Coaches and one from Players, resulted in a new, combined row of a new table.

The join operation works well with other operations to produce valuable query data. Suppose that a user of the data base system wanted a telephone list of the team coached by Henry Hazer. The following sequence of operations would produce such a list:

$$\pi_{\text{Player and Telephone}}(\sigma_{\text{Coach}=\text{Henry Hazer}}(\text{Coaches} \bowtie \text{Players}))$$

Given the table Coaches $\bowtie$ Players, the selection of coach = Henry Hazer would produce all 6-tuples with the *Coach* attribute value Henry Hazer. Of course, the value Breakers would be redundant in this case. Indeed, since only the names and telephone numbers of the players coached by Hazer are needed, the projection operation is the very ticket to these data.

Not surprisingly, the same information is often retrievable in many different ways from a relational data base. Not all ways yield the data at the same speed, however. For example, the join operation invoked in the sequence above is inherently inefficient because it is applied to two relatively large tables. An equivalent, but inherently more efficient, sequence of operations would be the following:

$$\pi_{\text{player and telephone}}(\text{Players} \bowtie (\sigma_{\text{coach}=\text{Henry Hazer}}(\text{Coaches})))$$

The selection operation with coach = Henry Hazer, when applied to Coaches, results in a single triple

Breakers, Henry Hazer, 731-5856

The join of Players with this triple thus specifies all players on the Breakers team. A final projection elminates all attribute values except *name* and *telephone* from the resulting 6-tuples.

This example illustrates an important feature of the kind of high-level query languages used with relational data bases and based on relational algebra. When fully implemented, such a language makes it possible for a user of such systems to optimize the queries, and given more than one way to skin a cat, one might as well do it quickly.

How is a relational data base implemented? Since only tables are maintained in the computer's memory, all relational operations must reduce to operations on tables. The selection operator is easily implemented. Simply scan the table, row by row, testing the boolean expression (that defines the rows to be selected) with attribute values from the row being scanned. Projection is also straightforward to program, namely by sorting the resulting rows and eliminating duplicates.

The join operation will run most quickly on two relatively large tables if the tables are first sorted in order of values for the common attribute. When the tables are sorted, they can be merged, via the common attribute, value by value.

There are many variations on the join operator. Instead of equality, the common attributes may be asked to satisfy other forms of comparison such as inequalities. Other operators besides projection, selection, and join are available. In most relational data base systems, queries can be formed on the basis of cartesian products, unions, intersections, and set subtraction.

In the 1970s, three major forms of data base system were introduced; the hierarchical system, the network system, and the relational system. It is now generally acknowledged that relational data bases have the widest applicability and usefulness. Consequently, only relational data bases are described in this book.

The languages in which queries in relational data bases are expressed have developed through several generations of sophistication. The queries illustrated so far in this chapter had an algebraic form and a procedural flavor. A recently developed query language called SQL is shaped more naturally for human use and is nonprocedural.

The basic query in SQL has the following general form:

SELECT $A_1, A_2, \ldots, A_n$
FROM $R_1, R_2, \ldots, R_m$
WHERE Boolean expression

Such a query implies not only the selection operation but also projection, cartesian product, and natural join. Arguments A_i specify which attributes are part of the query, and arguments R_j are relations. Users specify which relations they think will be required in producing an answer to the query. Indeed, the list R_1 through R_m defines a cartesian product of all the relations (tables) thus indicated. The boolean expression appearing in the "where" portion of the query includes a variety of operations, including selections and joins.

It can now be shown how some of the earlier queries framed in the language or relational algebra can be recast in SQL.

To produce a list of Breakers' players and their telephone numbers, one may write

```
SELECT   player, telephone
FROM     Players
WHERE    team = Breakers
```

The earlier example involving a list of players (and their telephone numbers) on the team coached by Henry Hazer would be obtained as follows in SQL:

```
SELECT   player, telephone
FROM     Players, Coaches
WHERE    Coaches.coach = Henry Hazer
         and Coaches.team = Players.team
```

Lest readers presume that SQL (or relational algebra, for that matter) is confined to relatively simple structures, the following example of a nested query in SQL is offered:

```
SELECT   player, telephone
FROM     Players
WHERE    team In
         (SELECT   team
          FROM     Coaches
          WHERE    coach = Henry Hazer)
```

SQL is currently becoming the industry standard query language for relational data base systems on large computers.

Problems

1. Given two tables with n rows, compare the worst-case times (see Chapter 14) taken by the following natural-join algorithms:
(*a*) Entry-by-entry comparison between the tables
(*b*) Sorting both tables on a common attribute, then merging
In both cases, assume that the common attribute takes n distinct values in each table.

2. Write a relational algebra expression that will result in lists of (*a*) players on the same team as Terence Tripper and (*b*) the dates and times of all Loons' home games.

References

Jeffrey D. Ullman. *Principles of Database Systems.* Computer Science Press, Rockville, Md., 1980.

C.J. Date. *An Introduction to Database Systems,* 3d ed. Addison-Wesley, Reading, Mass., 1981.

CHURCH'S THESIS

All Computers Are Created Equal

I
n 1936 the U.S. logician Alonzo Church formulated a sweeping thesis that expressed precisely what it means to compute. Church had struggled for some time with this notion, calling it *effective calculability*. It was to mean any process or procedure carried out stepwise by well-defined rules. Church believed that he had captured the notion precisely by means of a formal system called the λ calculus. His thesis, in precise terms, claimed that anything that might fairly be called effectively calculable could be embodied within the λ calculus.

When Church's paper appeared, there was already one other formal system that claimed to represent something similar, namely, the class of functions called *general recursive*. Within a year of Church's thesis, a third claimant appeared, namely, the Turing machine. An a priori judgment of the three paradigms of computing might result in a kind of Venn diagram, showing some overlap between the notions but some distinctions as well (Figure 66.1).

By what species of madness might one have supposed that all three notions would turn out to be the same? How was one to suspect that the functions computable by Turing machines were precisely the general recursive functions

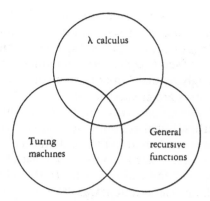

Figure 66.1 Three different notions of computing?

or that either notion (and no more) could be expressed within the λ calculus? Yet, Church's 1936 paper showed that the general recursive functions were precisely those framed by the λ calculus. Turing then demonstrated that the machine he had defined (see Chapter 31) was also equivalent to Church's λ calculus.

The λ calculus is a procedure for defining functions in terms of λ *expressions:*

A λ *expression* is an identifier, a string of λ expressions, or a number, or it has the form

λ (bound variable expression) $\cdot \lambda$ expression

A *bound-variable expression* is an identifier, the symbols ( ), or a list of identifiers.

Starting with some elementary λ expressions for numbers and operations on them, one may use this formalism (according to Church) to express any computable function.

The fact that the three notions of computability just mentioned turn out to be equivalent amounts to strong evidence in favor of Church's thesis. To be sure, there are models of computation such as the finite automaton (see Chapter 7) that turn out *not* to be equivalent to Turing machines, but only because they are less general. Church's thesis would appear to claim not only that all sufficiently general notions of computing will be equivalent, but also that there is a limit to that generality as well. Try as one might, there seems to be no way to define a mechanism of any type that computes more than a Turing machine is capable of computing. No general scheme will compute a function that is not recursive. No effective procedure falls outside the scope of the λ calculus.

Conversely, the failings of any one general scheme [such as the halting problem for Turing machines (see Chapter 59) is a failure for all. Church's thesis therefore puts a seemingly natural limit on what computers can do: All (sufficiently general) computers are created equal.

Support for Church's thesis comes not only from the equivalence of the three systems mentioned earlier, but also from other sources. For example, a fourth computational scheme is embodied in the random access machine (RAM, for short) (see Chapter 17). In Chapter 17 it was shown that a RAM can compute any function that a Turing machine can compute. In this chapter, we prove the opposite: Turing machines can mimic RAMs. The two proofs, taken together, establish the equivalence of the two concepts.

To show that a Turing machine can simulate a RAM, it is only necessary to represent the RAM's memory on a Turing machine tape and then to write a number of Turing machine programs, one for each possible RAM instruction. A RAM program can then be translated, instruction by instruction, to a Turing machine program that has the same effect.

In carrying out this translation, we encounter a problem with the unlimited word size permitted to RAMs. At first sight, it seems that the entire Turing machine tape will have to be devoted to a single RAM memory register! However, at any one time the RAM being simulated requires only a finite word size for any register currently in use. If we therefore devote just enough Turing machine tape cells to hold these contents, then a shift operation is implemented whenever the cells representing a given register are to be expanded; the Turing machine scans to a special marker indicating the end of the currently used memory cells. The Turing machine shifts the marker one cell to the right and does the same thing to each memory cell all the way back to the cells simulating the problem register. An extra cell is now available (see the problems).

Another labor-saving assumption involves the use of multiple tapes in the simulating Turing machine. It has already been shown (see Chapter 31) that each multitape Turing machine is equivalent to some one-tape machine. Figure 66.2 shows how a three-tape machine may simulate a RAM. The Turing machine may have any finite alphabet. It is therefore possible (as well as convenient) to let it use the ASCII alphabet in expressing the operation of a given RAM. The first tape used by the Turing machine contains the current contents of the accumulator. Also this tape has workspace for copying program instructions, computing new register contents, and doing similar things. The second tape is used to store the RAM's program, and the third tape stores the contents of the RAM's memory, with each register being represented by a number of consecutive tape cells.

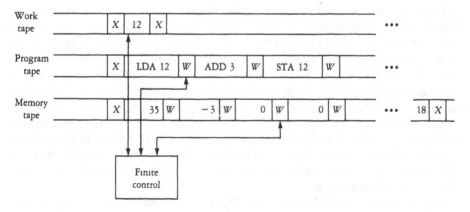

Figure 66.2 A Turing machine that mimics a RAM

Consider a single instruction in a hypothetical RAM program:

LDA 12

This means that the current contents of memory register 12 are to be transferred to memory register 0 (the accumulator).

Such an instruction will be recognized by the machine when tape head 2 scans the symbols LDA in sequence. It will then enter a subprogram that carries out the following steps:

1. Copy the operand (12) from tape 2 onto workspace in tape 1.
2. Move tape head 3 all the way to the left-hand *X* marker.
3. Scan rightward with tape head 3. Each time a *W* marker is passed, decrement the number stored in the workspace by 1.
4. When the workspace number reaches 0, copy all cells in tape 3 between the *W* markers into the cells simulating register 0.

Once these steps have been completed, control returns to the main program which then reads the next instruction code in order to decide what subprogram to enter next.

These initial conditions must be satisfied before the LDA (or any other instruction) subprogram can begin: Tape head 1 must scan the cell just to the right of the left-hand *X* marker, and tape head 2 must scan the leftmost cell of the operand for the instruction being processed. No particular condition applies to tape head 3. It can be anywhere.

State-transition diagrams give the simplest indication of just how the LDA

437

subprogram works. The four diagrams below correspond to the four steps listed earlier. Each circle represents a state and each arrow a transition. There is no need to label the circles except by the word *begin* appearing beside the state first entered in a given diagram. The arrows are labeled by three sets of symbols, including Greek letters. The first three symbols on an arrow represent the three symbols currently scanned by the three tape heads in the order 1, 2, 3. A dot means any symbol; a Greek letter means any symbol but one that must be written on another tape. An alphabetic letter means a transition triggered by the appearance on the corresponding tape of that specific character. In the second position, the three symbols represent the same things to be written. In the third position, motions of the three tape heads are referred to: L means left, R means right, and S means not at all.

1. Copy operand onto the work tape (Figure 66.3). The Turing machine's tape heads 1 and 2 move to the right, simultaneously copying the symbol σ that appears in tape 2 ($\cdot\sigma\cdot$) onto tape 1 ($\sigma\sigma\cdot$). This is done for any σ encountered except for the marker symbol W. When this symbol is encountered on tape 2, the program moves on to the next diagram, so to speak.

2. Move tape head 3 to the leftmost X (Figure 66.4). Here the Turing machine moves tape head 3 continually to the left, ignoring all symbols but the X when it finally arrives there. A new state is then entered. Here, tape head 3 is moved continually to the right with every symbol encountered on tape 3 being erased by a B. This device helps the Turing machine to find quickly the precise accumulator cell in which it must record a digit from memory register 12. The same idea was used in constructing certain components of the universal Turing machine (see Chapter 51). When the Turing machine encounters the register boundary marker W, the machine exits to the next diagram.

3. Scan rightward to "register" 12 (Figure 66.5). In the initial state of this diagram, the Turing machine scans rightward on tape 3 until it encounters a register boundary marker W. If there happens to be a nonzero digit σ currently under tape head 1, it is decremented and the Turing machine continues its rightward scan on tape 3. If tape head 1 currently scans a 0, however, it is replaced by a temporary A marker and the Turing machine then scans leftward on tape 1, skipping all 0s until it encounters the first nonzero symbol. If that

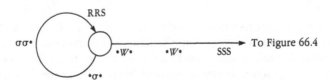

Figure 66.3

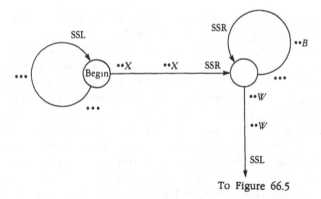

Figure 66.4

symbol is an *X*, then the number originally written on tape 1 has been decremented all the way to 0 and the machine exits to the next diagram. If the symbol is a nonzero digit, on the other hand, that digit is decremented and tape head 1 now moves to the right, converting every symbol up to and including the *A* to a 9. Now the machine reenters the first state in the diagram, looking for the next *W* marker.

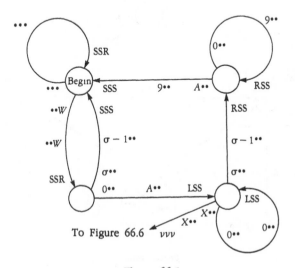

Figure 66.5

4. Copy register 12 into register 0 (Figure 66.6). When the Turing machine enters this diagram, tape head 3 scans the leftmost cell of the register strip specified by the LDA operand. In this case, it is the leftmost cell of register 12. It will be scanning one of digits 0 to 9 in such a case. And there are really 10 possible exits from the first state, but only two are shown in this diagram, one for 0 and the other for 9. In either case, the symbol is replaced by an A, and the next state entered is unique to the symbol just scanned. In this way, the Turing machine "remembers" the symbol without having to write it down. In the unique state, tape head 3 is continually moved leftward until a B is encountered. Readers will recall that register 0 was filled with B's in the second diagram. These are now replaced by the digits occupying register 12, one digit at a time in right-to-left fashion. Thus when the Turing machine enters the third state in the diagram above, it replaces the B by the remembered digit and then moves rightward along tape 3 until the A symbols are encountered again. It continues scanning rightward until it discovers either a digit or a W marker. In the first case, it proceeds to copy that digit into the 0 register strip as before. In the second case, it exits from the diagram.

In leaving this diagram, the LDA instruction has been executed: The contents

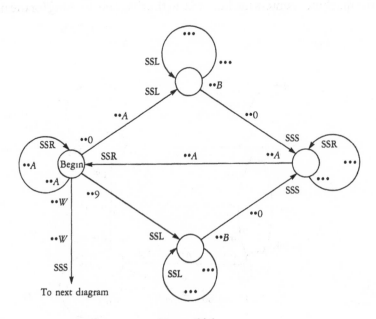

Figure 66.6

of register 12 have been transferred to register 0. An additional diagram, not shown, suffices to put tape heads 1 and 2 back in their proper initial positions.

The shift operation was not required in the LDA subprogram, but it is required by ADD, SUB, STA, and STI (see Chapter 48). At certain critical points in these operations, the Turing machine may require one more cell than is available in the current 0 register strip, so the Turing machine suspends current operation and invokes the shift subprogram. When this happens, the Turing machine alternates between two states. This many states are sufficient to remember each symbol on tape 3 and then to write it one time step later, provided that the scratch tape (1) is used properly.

This completes the description of just one of the 12 subprograms needed by a Turing machine to simulate RAMs. The total program is quite large, but this is not surprising in view of the fact that Turing machines are conceptually much simpler than RAMs. In any event, the full construction proves that a Turing machine can carry out any computation that a RAM can—another piece of evidence in favor of Church's thesis.

Problems

1. Draw a state-transition diagram for the shift subprogram mentioned in the text. From a given cell on tape 3, it shifts the contents of every cell out to the right-hand X marker one position to the right.

2. Given an effective shift subprogram, write a subprogram for the STA instruction.

References

S. C. Kleene. *Introduction to Metamathematics.* Van Nostrand, Princeton, N.J., 1960.

Derek Wood. *Theory of Computation.* Harper and Row, New York, 1987.

INDEX

453